# Divided Brains
The Biology and Behaviour of Brain Asymmetries

Asymmetry of the brain and behaviour (lateralization) has traditionally been considered to be unique to humans. However, research has shown that this phenomenon is widespread throughout the vertebrate kingdom and is found even in some invertebrate species. A similar basic plan of organization exists across vertebrates.

Summarizing the evidence and highlighting research from the past 20 years, the authors discuss lateralization from four perspectives – function, evolution, development and causation – covering a wide range of animals, including humans. The evolution of lateralization is traced from our earliest ancestors, through fish and reptiles to birds and mammals. The benefits of having a divided brain are discussed, as well as the influence of experience on its development. A final chapter discusses outstanding problems and areas for further investigation.

Experts in this field, the authors present the latest scientific knowledge clearly and engagingly, making this book a valuable tool for anyone interested in the biology and behaviour of brain asymmetries.

**Lesley J. Rogers** is Emeritus Professor at the Centre for Neuroscience and Animal Behaviour, University of New England, Armidale, Australia. A Fellow of the Australian Academy of Science, she has made outstanding contributions to understanding brain development and behaviour, including the discovery of lateralization in the chick forebrain at a time when lateralization was thought to be unique to humans.

**Giorgio Vallortigara** is Professor of Neuroscience at, and Director of, the Centre for Mind/Brain Sciences, University of Trento, Rovereto, Italy. His research includes the study of spatial cognition in the avian brain, number and object cognition in animals and lateralization of cognition. He discovered functional brain asymmetry in the so-called 'lower' vertebrate species.

**Richard J. Andrew** is Emeritus Professor at the School of Life Sciences, University of Sussex, Brighton, UK. He has worked extensively on lateralized processes in memory formation in chicks, and on behavioural transitions during early development. At present he uses zebrafish to explore the role of brain asymmetries in the generation of lateralized behaviour.

# Divided Brains

## The Biology and Behaviour of Brain Asymmetries

### Lesley J. Rogers

University of New England,
Armidale,
Australia

### Giorgio Vallortigara

University of Trento,
Rovereto,
Italy

### Richard J. Andrew

University of Sussex,
Brighton,
UK

CAMBRIDGE
UNIVERSITY PRESS

# CAMBRIDGE
## UNIVERSITY PRESS

Shaftesbury Road, Cambridge CB2 8EA, United Kingdom

One Liberty Plaza, 20th Floor, New York, NY 10006, USA

477 Williamstown Road, Port Melbourne, VIC 3207, Australia

314–321, 3rd Floor, Plot 3, Splendor Forum, Jasola District Centre, New Delhi – 110025, India

103 Penang Road, #05–06/07, Visioncrest Commercial, Singapore 238467

Cambridge University Press is part of Cambridge University Press & Assessment, a department of the University of Cambridge.

We share the University's mission to contribute to society through the pursuit of education, learning and research at the highest international levels of excellence.

www.cambridge.org
Information on this title: www.cambridge.org/9781107005358

First published 2013

*A catalogue record for this publication is available from the British Library*

*Library of Congress Cataloging-in-Publication data*
Rogers, Lesley J.
Divided brains : the biology and behaviour of brain asymmetries / Lesley J. Rogers, Giorgio Vallortigara, Richard J. Andrew.
    p.   cm.
ISBN 978-1-107-00535-8 (hardback)
1. Cerebral dominance.   2. Brain – Duality.   3. Brain – Anatomy.   I. Vallortigara, Giorgio, 1959–   II. Andrew, Richard John, 1932–   III. Title.
QP385.5.R64   2013
612.8′2–dc23

                                                                    2012023177

ISBN   978-1-107-00535-8   Hardback
ISBN   978-0-521-18304-8   Paperback

# Contents

| | | |
|---|---|---|
| *List of illustrations* | | *page* vi |
| *Advance praise* | | viii |
| *Preface* | | ix |
| 1 | Introduction | 1 |
| 2 | Function | 35 |
| 3 | Evolution | 62 |
| 4 | Development | 98 |
| 5 | Causation | 123 |
| 6 | Applications and future directions | 153 |
| *References* | | 172 |
| *Index* | | 218 |

# Illustrations

| | | |
|---|---|---|
| 1.1 | Habenular asymmetry in the convict cichlid fish | *page 2* |
| 1.2 | Monocular testing | 7 |
| 1.3 | Frequency distributions of asymmetry | 11 |
| 1.4 | The pebble floor task | 14 |
| 1.5 | Tail wagging measurements | 16 |
| 1.6 | Bottom-up and top-down inputs to the nucleus rotundus in the pigeon | 19 |
| 1.7 | A 'cancellation task' | 21 |
| 1.8 | Summary of left–right differences | 28 |
| 1.9 | Lateralization in bees | 29 |
| 2.1 | Crossing of nerves | 36 |
| 2.2 | Wiring of sensory motor units | 37 |
| 2.3 | The C-start reaction of a fish | 40 |
| 2.4 | Spatial behaviour | 43 |
| 2.5 | Dual task | 45 |
| 2.6 | 'Termite fishing' by chimpanzees | 49 |
| 2.7 | Game-theoretical model of left–right proportions in a group | 53 |
| 2.8 | Antenna responsiveness of bees | 59 |
| 2.9 | Laterality of Tanganyikan scale-eating cichlid | 60 |
| 3.1 | *Branchiostoma*, a lancelet | 63 |
| 3.2 | Phyletic tree | 65 |
| 3.3 | *Haikouella* fossils | 66 |
| 3.4 | A Conodont | 69 |
| 3.5 | Parietal eye of a lizard | 71 |
| 3.6 | Human brain | 81 |
| 3.7 | The slug *Triboniophorus* | 91 |
| 3.8 | *Histioteuthis*, a deep-sea squid | 93 |
| 4.1 | Chick embryos at two stages of development | 103 |

4.2   The visual pathways of birds                                      105
4.3   Attack and copulation                                            106
4.4   Visual pathways of the pigeon                                     108
4.5   An Australian brush turkey, *Alectura lathami*                    110
4.6   Nest of the Brazilian rufous hornero (*Furnarius rufus*)         112
4.7   Eggs of three species                                            113
4.8   Shifts in hemispheric dominance                                  120
5.1   Sniffing dog                                                     125
5.2   Australian magpie, *Gymnorhina tibicen*, looking overhead        141
5.3   Viewing a predator                                               142
6.1   Horses attacking                                                 155
6.2   Testing lateralized auditory responses in dogs                   156
6.3   Portrait of a gentleman                                          169

# Advance praise for *Divided Brains*

"This fascinating book has been written by three experts in the field. The different roles played by the two sides of the brain were thought to be a uniquely human characteristic, but the authors show that such lateralisation has ancient origins in biological evolution. They have written a superb book which I shall use as an invaluable source for years to come."

Professor Sir Patrick Bateson, University of Cambridge, co-author of *Plasticity, Robustness, Development and Evolution* (Cambridge, 2011)

"Birds do it, bees do it – and so, it seems, do species of every taxa: They show cerebral and behavioral asymmetries that belie the seeming bilateral symmetry of the body, and even the brain itself. Until quite recently such asymmetries, especially in the form of right-handedness and left-brain dominance, were held to be uniquely human, and even to define our species. This anthropocentric view is here comprehensively buried. The book is more than simply a compendium of asymmetries across different species. Rogers, Vallortigara and Andrew cover evolutionary, development and genetic aspects of asymmetry, asking why and how asymmetries evolved in a world that is indifferent to left and right. This is the most in-depth analysis to date, by the three foremost authorities on animal asymmetries, of a phenomenon that has fascinated scientists and philosophers through the centuries."

Professor Michael C. Corballis, University of Auckland

"A timely addition to our understanding of hemisphere difference, this book is a vital and accessible source of information about laterality in fish, reptiles, birds, mammals, and even insects. It does not content itself with merely marshalling information, though it does that very well, but addresses the 'how' and 'why' of the asymmetrical world of all living things."

Dr Iain McGilchrist, author of *The Master and his Emissary: The Divided Brain and the Making of the Western World*

"In the last 30 years it has become clearer and clearer that there are functional differences between the two sides of the brain in vertebrates and even in invertebrates, and that these differences sometimes reveal deep phylogenetic trends. It is unlikely that any other group of authors could have done such a remarkable synthesis of the current state of evidence on this topic."

Professor Peter F. MacNeilage, Department of Psychology, University of Texas at Austin

# Preface

Research on lateralization of brain and behaviour in animals has expanded rapidly over the past two decades and continues to grow exponentially. The same is true of studies on lateralization in humans, and the evidence from these two sources is integrated in this book in a way not previously attempted. We were motivated to write this book because of the widening interest in the subject and a perceived need to make the most up-to-date information available in a form that, we hope, is stimulating and easy to read. Since there are many general texts on cerebral specialization in humans, our chief focus was on left–right differences in brain and behaviour in non-human animals, with the aim of bringing together recent striking advances arising from study of lateralization in these species and the state of knowledge of lateralization in humans.

We approached the topic of lateralization from the perspective of Tinbergen's four questions (function, evolution, development, causation), to each of which we have devoted one chapter.

# Introduction

*Once the emperor Hui Tsung was enjoying the sight of a lichee tree laden with fruit before the palace when a peacock approached the tree, and he summoned his artists at once to make a picture. They produced a magnificent painting of the peacock with its right foot poised to take a step on a flower-bed: but to their surprise the emperor shook his head over it. A few days later when he asked if they had discovered their mistake, they had no answer ready. Then Hui Tsung told them: 'A peacock always raises its left foot first to climb.'*

Cheng Chen-To, Chang Heng and Hsu Pang-Ta (1957)

## Summary

Once thought to be unique to the human brain, lateralization of structure and behaviour is now known to be widespread in vertebrates and, furthermore, it has a similar plan of organization in the different species. This chapter introduces the basic pattern of lateralization of vertebrate species and does so in a historical context to highlight the fact that, until some 20 years ago, it was widely and incorrectly assumed that having a lateralized brain was a mark of the cognitive superiority of humans. It also introduces some of the new evidence showing the presence of lateralization in invertebrate species.

## 1.1 Introduction

It is difficult to understand why incorrect ideas in science sometimes establish such deep roots that it is very difficult to eradicate them. Paul Broca, the French physician and anthropologist, is numbered among the founders of research on brain asymmetry. Discussing left–right differences in the brain, in 1865 he wrote, 'there is a less but still very evident degree of dysymmetry in the great apes' (Broca, 1865, p. 527). Judging from the excerpt above on the emperor and the peacock (see also Humphrey, 1998) and from notes on foot preferences in parrots (Harris, 1989), the existence of asymmetry in the behaviour of animals had been acknowledged for a very long time, but outside academic circles. Until relatively recent years, however, textbooks of neuroscience and psychology referred to brain asymmetry as a uniquely human attribute, linked to superior cognitive abilities of our species and, in particular, to language.

Maybe this omission should be considered a late outcome of a mistaken concept of evolution based on the ancient idea of *Scala Naturae* (see Hodos and Campbell, 1969). Yet, even assuming that certain phenomena such as the handedness so conspicuous in human behaviour were not so clear in other animals (dismissing for the moment that this could have been simply because other species do not have or do not use appendages in any way similar to that of humans), the raw facts of anatomy should have been difficult to dismiss. Consider the evidence for anatomical asymmetries at the level of the midbrain or diencephalon (which we shall discuss in detail in Chapters 3 and 5). Diencephalic asymmetries were common knowledge among neuroanatomists at the beginning of last century (see Braitenberg and Kemali, 1970). For instance, habenular asymmetries, referring to a collection of cells in the dorsal thalamus of the brain, had been observed in the most primitive living verte-brates, the jawless fish (Cyclostomes), e.g. the lamprey (see Braitenberg and Kemali, 1970). However, any mention of these asymmetries subsequently disappeared from anatomy textbooks, and they have been re-discovered only

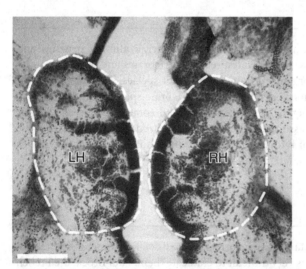

Figure 1.1 Habenular asymmetry in the convict cichlid fish. Brightfield microphotograph of a coronal section through the habenula of a female convict cichlid (*Amatitlania nigrofasciata*). In this individual, the left habenula is 28.01% larger in volume than the right (calculated by measuring areas on many sections). The laterality index is –0.123. Scale bar = 200 μm. Microphotograph courtesy of Professors Peter L. Hurd and Cristian Gutierrez.

recently. An example is shown in Figure 1.1, which depicts habenular asymmetry in the convict cichlid fish *Amatitlania nigrofasciata*, where the left habenula is larger than the right (Gutiérrez-Ibánezfoun *et al.*, 2011).

Turning from brain structure to function, we must wait until the late 1970s or early 1980s for the first evidence of functional lateralization of the animal brain. Perhaps as an illustration of the operating of the Zeitgeist, more or less simultaneously Fernando Nottebohm (1971, 1977) reported asymmetrical control of song production in two species of songbird; Lesley Rogers (Rogers and Anson, 1979) reported functional asymmetries in domestic chicks; and Victor Denenberg (1981) reported asymmetries in rats. The techniques used to reveal these asymmetries were severing of nerves that control singing, injection of a pharmacological agent into one or the other hemisphere of the brain and placement of unilateral lesions on the left or right side.

Nottebohm's findings were probably the most well known outside the realms of neurobiologists specializing in avian and rodent anatomy and behaviour, because they seemed to provide a direct link to lateralization of language in the human brain (see Chapter 5). Cutting of the branch of the left hypoglossal nerve, which innervates the muscles on the left side of the syrinx, the organ producing song in birds, severely impairs the bird's ability to produce song, whereas lesions of the equivalent nerve on the right side have no effect on song. Similar results are produced by lesions of the left, but not the right, higher vocal centres in the brain (Nottebohm *et al.*, 1976; Nottebohm, 1977, 1980).

More recent research has confirmed these findings, although species differences seem to exist (Schmidt *et al.*, 2004). Moreover, hemispheric specialization has been observed for perception, rather than production, of song by passerine birds. George *et al.* (2004) recorded neuronal responses in the primary auditory area of the songbird brain, the Field L complex, to species-specific and artificial sounds in both awake and anaesthetized male starlings (*Sturnus vulgaris*). They found significantly more responsive neurons in the right hemisphere than in the left hemisphere of awake birds, and this difference was significantly reduced in anaesthetized birds. Clear hemispheric specialization towards categories of behaviorally relevant stimuli and precise parameters of these stimuli were found in awake birds: the right hemisphere responded most strongly to species-specific sounds, particularly to familiar vocalizations and bird's own individual-specific whistles, and the left hemisphere responded to unfamiliar individual-specific songs. When the birds were anaesthetized, the left hemisphere responded more than the right to artificial, non-specific stimuli. Furthermore, it is known that songbirds are able to discriminate between their own song and the songs of conspecifics. Using functional magnetic resonance imaging (fMRI), Poirier *et al.* (2009) showed that this selectivity is present at midbrain level in adult male zebra finches (*Taenopygia guttata*) and lateralized towards the right side.

It has been argued that control of speech in humans is made possible by two distinct mechanisms: a feedback control mechanism, by which speech production is monitored during speaking and any deviation from the expected signal is corrected on the basis of auditory information, and a feed-forward control mechanism, by which speech is produced on the basis of previously learned commands, without reliance on incoming sensory information (Guenther, 2006). Tourville *et al.* (2008) suggested that, while the left lateralization of speech production deduced from studies using lesions would reflect left lateralization of the feed-forward control system, right lateralization of the auditory feedback control would explain the importance of the right hemisphere observed in numerous aspects of speech production, such as self-recognition processes (Fu *et al.*, 2006). The latter seem to be lateralized to the right hemisphere in both the auditory (Rosa *et al.*, 2008) and the visual modality (Keenan *et al.*, 2001). It is possible that the right lateralization to discriminate between the bird's own song and other conspecific songs, as found in zebra finches (see above), is due to a right lateralization of the auditory feedback control system, suggesting important anatomical and functional similarities between birds and humans.

Discovery of lateralization in domestic chickens (Rogers and Anson, 1979) is interesting because, in contrast to other examples, such asymmetry was not actively searched for but rather observed serendipitously. The same was the case in later research on domestic chicks: for example, lateralization of imprinting memory constituted a nuisance for research on the biological bases of memory.[1]

Intracranial injection of substances that interfere with memory consolidation was used widely as a tool for investigation of memory consolidation and the domestic chick as an animal model for such research (see for reviews Andrew, 1991b). At the time it was implicitly assumed that the site of the injection, to the left or to the right hemisphere, would be immaterial. Lesley Rogers, however, proved for the first time that injecting cycloheximide (an inhibitor of protein synthesis) into the chick's left hemisphere produced distinct effects on visual discrimination learning and auditory habituation that were absent when the injection was performed in the right hemisphere (Rogers and Anson, 1979).

In a similar vein, in the early 1970s a research group formed by the ethologist Patrick Bateson, the anatomist and neurobiologist Gabiel Horn and the neurochemist Steven Rose started to use the phenomenon of filial imprinting (a process by which young precocial, nidifugous birds come to recognize their mother and social partners by being exposed to them briefly) in the domestic chick as a model-system for investigation of the biological bases of memory (Horn *et al.*, 1973). After identification of a plausible area of the brain as a putative candidate

---

[1] A remarkable comment by neurochemist Steven Rose on Sir John Eccles's claims about lateralization in humans may serve as advice to those who, with little mastering of biological literature, would still argue about human uniqueness as to brain asymmetry: 'But if Eccles did turn out to be right, and functional lateralization is the key to possession of a soul, then any of my chicks would have as good claims as Sir John to possessing one' (Rose, 1992, p. 249). Eccles could be excused at the time, but more recent epigones cannot be so.

for the site of storage of imprinting memories, the intermediate medial meso-pallium (IMM, an associative area of the chick forebrain), researchers started a series of control experiments to disentangle specific effects of learning from non-specific brain activation related to sensory and motor activity associated with exposure to the imprinting stimulus. The assumption was that imprinting pro-cesses in the left or the right hemisphere would be the same. However, results from sequential lesioning experiments at various times after imprinting and analyses of plastic changes at the synapses soon revealed an unexpected pattern of lateralization. Evidence suggested that both the right and the left IMM act as short-term memory stores, but only the left IMM is used as a long-term store (Cipolla-Neto et al., 1982). The right IMM is crucial in establishing another store, somewhere outside the IMM region, referred to as S′. The right IMM passes information on to S′ over a period of several hours. It has been suggested that passing of the memory from the right IMM to S′ may add to the depth of processing by allowing the storage of contextual information and thus enriching simple representations initially stored in the IMM (see Horn, 1985, 2004).

Another experimental paradigm mostly used in research on the biological aspects of memory formation is the so-called passive avoidance learning (PAL) task, originally introduced by Cherkin (1969) to test chicks. The standard version of the task involves the presentation of a coloured (e.g. red) bead, at which chicks will readily peck, coated with a bitter-tasting substance. After this training, chicks will subsequently avoid pecking at a bead of similar colour and size (but not at a bead of a different colour, e.g. blue) (Lössner and Rose, 1983). Long-lasting learning occurs after a single and sharply timed experience (peck-ing the bitter bead), enabling scientists to study the time course of memory formation with great precision. The formation of a memory of the PAL task occurs over the course of hours, with a range of well-documented biochemical, physiological and morphological changes occurring mainly in the intermediate medial mesopallium, but also in structures such as the medial striatum, StM (Rose, 2000). In the hours following training the memory trace becomes fragmented and redistributed in different structures. In particular, circuits in the IMM might retain some aspects of the memory trace (e.g. the colour of the bead), whereas other aspects (e.g. the size and shape of the bead) might be encoded by the StM (Rose, 2000). The changes observed at different levels in chick's forebrain, after the training experience, are associated with different memory phases (short-term, intermediate-term and long-term memory), defined on the basis of sharply timed on/offsets of sensitivity to different amnesic agents and memory loss at specific times after training (e.g. Gibbs et al., 2003), as well as brief enhancements of memory recall (Andrew, 2002a).

While studying the time course of memory formation, the presence of struc-tural and functional lateralization was noticed. Evidence seemed to indicate that the memory for the standard PAL task forms mainly in the left hemisphere. A seminal finding was that bilateral or left, but not right, lesions of the mesopallium made before training resulted in interference with acquisition of the task (Patterson *et al.*, 1990). Moreover, unilateral injections of the amnesic agents used to determine memory phases revealed that, in most cases, the timing of the effects of left hemisphere injections was identical with that of bilateral injections (Gibbs *et al.*, 2003). The trace encoded by the left hemisphere is, therefore, considered to be largely responsible for subsequent performance and for the processes involved in the phases of memory formation, since injections of amnesic agents into the right hemisphere are usually ineffective (Gibbs *et al.*, 2003). In particular, a crucial left hemispheric involvement might be prevalent in the earlier stages of memory formation, with participation of the right hemisphere in later encoding (e.g. Rickard and Gibbs, 2003a, 2003b), such as during the intermediate-term memory phase (Gibbs *et al.*, 2003). This is also consistent with biochemical evidence showing that the memory trace appears to consolidate first in the left mesopallium and then in the right (Sandi *et al.*, 1993).

The involvement of the right hemisphere in intermediate-term memory (manifested in a transitory susceptibility to amnesic agents during that phase) suggested that there is normally an interaction between left and right hemispheres at this stage (even though the left hemisphere seems to be still the dominant one). A putative function of this sort of interaction would be that of establishing linkages between the memory traces held in the two hemispheres, each encoding different aspects of the same experience (Andrew, 1997, 1999). Thus, the successful consolidation of memory traces would depend on the integration of information about the learning task encoded in both hemispheres. When the trace of the right hemisphere is degraded, due to the effect of amnesic agents, the interaction between the two hemispheres leads to a decrease of performance.

The predominant role of the left hemisphere in memory formation for the standard PAL task is likely to be due to its importance for the control of motor 'manipulative' responses towards objects, including those performed with the chick's beak (Andrew *et al.*, 2000). In addition, the left hemisphere has a role in the discrimination of local, specific cues associated with a target (such as the colour of the bitter-tasting bead that allows the chick to tell it apart from a neutral bead; see Vallortigara *et al.*, 1996; Tommasi and Vallortigara, 2001). Whereas the information stored by the left hemisphere is about the properties of objects to be manipulated, the detailed representation encoded in the right hemisphere involves mainly elements such as position and spatial context (as will be discussed in the next sections).

The possibility of establishing use of the chick as an ideal experimental model for the study of brain lateralization was then complemented by the first experimental evidence of visual lateralization in an intact non-human animal. This was shown by Richard Andrew using temporary occlusion of either the left or right eye (Andrew *et al.*, 1982; Figure 1.2), which takes advantage of the virtually complete decussation (crossing over the midline) of optic nerve fibres at the optic chiasma in birds with laterally placed eyes.

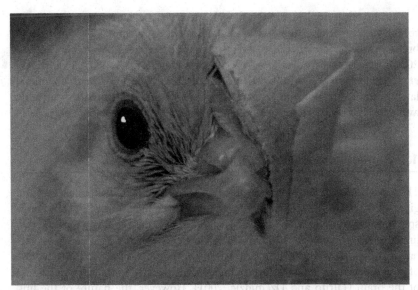

Figure 1.2 Monocular testing. A chick wearing an eye-patch for behavioural tests of brain lateralization. Photograph courtesy of Dr Cinzia Chiandetti, laboratory of Professor Giorgio Vallortigara.

Andrew *et al.* (1982) showed that the right eye was better at discriminating visual stimuli, such as grains from pebbles, and that the left was more reactive to emotionally charged stimuli (note that this result is consistent with the findings of Rogers and Anson, 1979). Charles Hamilton, a former pupil of Nobel Prize winner for research on split-brain patients (Roger Sperry), commented: 'These results lead to the plausible but revolutionary inference that a bird more effectively searches for food with its right eye while it watches for danger with its left!' (Hamilton, 1988). The results were duplicated and extended to several other species, and the basic pattern of hemispheric specialization confirmed (reviewed by Vallortigara, 2000; Rogers, 2002a; Rogers and Andrew, 2002).

The field of research on lateralization in animal models is nowadays huge and rapidly expanding. Although studies of lateralization in non-human species is clearly at odds with the idea that cerebral asymmetry is a unique characteristic of humans, there could be specifically human abilities that are also lateralized (for example this is likely to be the case in aspects of language). However, although evidence of lateralization in non-human animals is now penetrating into textbooks (e.g. Breedlove *et al.*, 2010), we are not persuaded that neuroscientists have a clear perception of how widespread it is in non-human species. The reason is that the literature reporting lateralization is sparsely distributed in a variety of journals, mostly of ethology, behavioural biology and animal neuroscience. This state of affairs provided an impetus for writing this book.

The aim of the book and of this first chapter in particular is not to provide the reader with an exhaustive review of the evidence for animal asymmetries in brain and behaviour, because several specialized reviews cover this (Rogers, 2002a; Rogers and Andrew, 2002; Vallortigara and Rogers, 2005; Vallortigara et al., 2011). Here we want to provide a general overview of research on animal lateralization, showing the variety of species, methods and findings currently available, and trying to make a little sense of it.

## 1.2 Handedness and other motor asymmetries

The most notable example of lateralization in humans is right-handedness, and given that both language and right-handedness are functions of the left hemisphere (e.g. Santrock, 2008), it has been argued that they could be in some way linked (e.g. Broca, 1865; Hellige, 1993a, 1993b). Humans exhibit 90% right-handedness (McManus, 2002) and within this population approximately 95% of individuals have language-processing regions situated in the left hemisphere of the brain (Lurito and Dzemidzic, 2001). However, the nature of the link between right-handedness and language is hotly debated (Corballis, 2002, 2003; Vauclair, 2004).

It has long been denied that non-human animals show differences in the use of their limbs in any way comparable to human handedness, and early research on great apes, our closest relatives, seemed to confirm this view. Historically, with only one exception (Boleda et al., 1975), results suggested that great apes did not express a right-hand population bias similar to humans (e.g. Finch, 1941; Marchant and Steklis, 1986). However, this view has been completely changed by a re-analysis of the data on hand preferences in primates by MacNeilage et al. (1987) and more recent systematic investigations with large sample sizes. Right-hand biases in great apes have been reported in captive chimpanzees (Pan troglodytes), related to complex manual tasks such as bimanual feeding, coordinated bimanual actions, bipedal reaching and throwing (for reviews, see Hopkins 2006, 2007), and in captive gorillas (Gorilla gorilla) (Byrne and Byrne, 1991; Meguerditchian et al., 2010a) for bimanual coordinated actions. Criticisms were raised about these initial reports, arguing that they were based on single laboratory samples (reviewed by McGrew and Marchant, 1997; Papademetriou et al., 2005), or on methodological and theoretical grounds (McGrew and Marchant, 1997; Palmer, 2002, 2003) and the suggestion that apes' exposure to human culture might have induced a bias of hand use in manual actions (e.g. McGrew and Marchant, 1997). However, such criticisms appear to be untenable since new data in support of a right-hand bias continues to mount from an increasing number of great ape species for a range of manual actions (e.g. Hopkins et al., 2004; Llorente et al., 2009; Meguerditchian et al., 2010a, 2010b; Llorente et al., 2011) across both captive and wild settings (Lonsdorf and Hopkins, 2005;

Llorente *et al.* 2011). Morcover, the controversy concerning whether handedness is observed only in apes in captivity, and therefore is possibly an artefact of apes imitating the right-handedness of their human caretakers, was dissolved recently by a large meta-analysis of studies on 1524 great apes, which revealed that right-handedness in chimpanzees and bonobos manifests itself irrespective of rearing conditions (Hopkins, 2006).

It has been argued, however, that the departure from a random distribution in these animal populations is typically small, compared with handedness in humans. The ratio of right- to left-handed chimpanzees is about 2:1 or 3:1 (in the case of gesturing and throwing), which is lower than most reports of handedness in various human cultures (Annett, 2006). This argument, however, should be taken with caution. Although right-handedness in humans appears to be a robust and universal finding (Perelle and Ehrman, 1994; Raymond and Pontier, 2004) the evidence that supports a 90% right-handed and 10% left-handed population split is mainly derived from self-report questionnaires in literate populations (e.g. Oldfield, 1971; Hardyck *et al.*, 1975; McManus, 1981). Although questionnaires rely primarily on measures of precision tool use, handedness patterns become more complex when a more ethological range of factors is considered, and right-handedness can then vary between 70 and 90% (Annett, 2002). Marchant *et al.* (1995) tracked naturalistic handedness across three different pre-literate populations and noted that, while there was an overall consistent but rather weak right-hand dominance (about 45:55 for left:right), individuals were mixed-handed for all actions across a comprehensive range of ethological measures with the exception of tool use, which was distinctly right-handed. Further studies in traditional cultures have shown that the percentage of left-handedness fluctuates widely (3–27%) (Faurie and Raymond, 2005; see Chapter 2). Most important, it should be considered that 70% of left-handers still exhibit a left hemisphere dominance for language functions (e.g. Knecht *et al.*, 2000), thus calling into question that a bias in handedness represents a reliable marker of hemispheric specialization for language. Vauclair and Meguerditchian (2008) argued that the dominant hand for manual gesture may constitute a more accurate marker of language lateralized hemisphere.

Therefore, it remains to be seen whether the distribution of handedness differs between apes and humans and, if it does, whether it reflects the emergence of socio-cultural evolution or alterations in the genome between great apes and humans. The fact that the distribution of preferences in limb use in other species may conform to or be even stronger than that of humans would argue against the second possibility. Parrots and cockatoos that use their feet to manipulate food and objects with a high degree of sophistication have significant footedness present at the population level with proportions similar to those of handedness of precision-gripping tool use in humans (Rogers, 1980; Harris, 1989; Rogers and Workman, 1993).

Neuroanatomical studies have indicated that all four species of great apes display regions of the brain homologous to the speech and language areas in the human brain – Broca's area (Cantalupo et al., 2003) and Wernicke's area (Spocter et al., 2010) – and they are larger in the left than in the right hemisphere (see Figure 3.6). In chimpanzees a leftward bias in cortical gyrification is present in right-handed animals, whereas it is absent in non-right-handed animals (Hopkins et al., 2007). In the primary motor cortex of chimpanzees a higher neuronal density of layer II/III cells on the left side has been documented (Sherwood et al., 2007).

Interestingly, measures of unimanual actions directed towards target objects (animate: self, social partner; inanimate: object, environment, enclosure) in semi-naturalistic conditions revealed in both gorillas (Forrester et al., 2011) and chimpanzees (Forrester et al., 2012) a significant right-hand bias for actions directed towards inanimate targets and no significant preference for use of the left or right hand for actions directed towards animate targets. The results may reflect the differential processing capabilities of the left and right hemispheres, as influenced by the emotive (animate) and/or functional (inanimate) characteristics of the target, respectively. Forrester and colleagues speculate that right-handed hierarchical object manipulation may have served as a precursor to modern human language skills. In great apes, communicative gestures may represent an evolutionary step towards language skills, extending the left hemisphere's specialized processing derived from tool use.

Evidence of handedness is also apparent in monkeys, as first determined by MacNeilage et al. (1987). Similar to chimpanzees (and humans), some species of monkeys show evidence of only individual-level handedness in simple behavioural tasks (e.g. marmosets; Hook and Rogers, 2008) but some species are lateralized at the population level in more complex tasks (Fagot and Vauclair, 1991). Baboons prefer to use the left hand during fine motor-spatial tasks such as object alignment, haptic discrimination, catching live fish or joystick manipulation, but prefer to use the right hand when they have to extract food from a narrow tube or gesture towards other monkeys (e.g. Meguerditchian and Vauclair, 2006). Humans also seem to be better at fine adjustments and haptic discrimination with the left hand (e.g. Fagot et al., 1997) and at fine motor tasks and gesturing with the right hand. As in the case of chimpanzees, a link between hand preference and structural brain asymmetry has been reported in monkeys (see for example in marmosets, Gorrie et al., 2008).

In non-primate mammals it was also once believed that limb preferences existed only at the individual but not population level (Figure 1.3 – the former referring to the case in which individuals are lateralized but with a 50:50 distribution of left- and right-hand biased individuals; the latter to the case in which the majority of the individuals in the population shows preference in a particular direction, e.g. right-hand bias in humans; see Chapter 2). However, more recent work has proved that this is not true: large samples of inbred mice

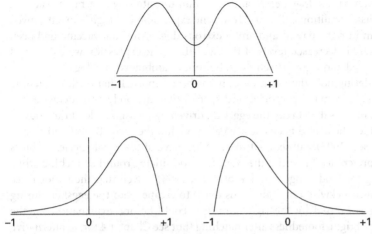

Figure 1.3 Frequency distributions of asymmetry. Top: individual-level asymmetry (with a similar number of left- and right-asymmetric individuals). Bottom: population-level asymmetry (most individuals being right- or left-asymmetric). −1 indicates all left preferring, 0 no asymmetry and +1 all right preferring.

have shown significant right-pawedness in tests of lateral paw preference and left-pawedness in reaching tests (Waters and Denenberg, 1994). Right paw preference at the population level has also been found in rats (73% of the population; Güven et al., 2003).

Dogs also show paw preferences at the population level, though varying with sex: males show a clear left-paw preference, whereas females show only a trend to prefer the right paw (Quaranta et al., 2004). The reason for this is, at present, unclear. However, it seems to be consistent with previous evidence by Tan (1987) reporting a slight preference for right paw use in a sample of dogs in which females were the larger group (19 out of 28 animals tested), and a similar pattern of sex modulation of paw preferences has been observed in cats (Wells and Millsopp, 2009).

Dogs' pawedness is also linked to different patterns of immune responses (Quaranta et al., 2004). It has been shown that the titres of specific IgG (immunoglobulin G) anti-rabies antibodies after immunization with rabies vaccine were lower in left-pawed dogs than in right-pawed or ambidextrous dogs, as determined in a task requiring the dog to remove a piece of adhesive tape from the snout. Similarly, the IFN (interferon)-γ serum levels were found to be lower in left-pawed dogs than in right-pawed or ambidextrous dogs. Mice with left-turning preferences during spontaneous rotational behaviour in a circular cage also show lower serum IFN-γ than mice with right-turning preferences (Kim et al., 1999). Quaranta et al. (2008) measured the expression of interleukin 2 (IL-2) and interleukin 6 (IL-6) genes in left-pawed, right-pawed and ambidextrous dogs before and after immunization with a rabies vaccine. Under basal conditions, IL-2 and IL-6 gene expression was higher in left-pawed dogs than in right-pawed and ambidextrous dogs. After the vaccine had been administered, decreased levels of IL-2 and IL-6 gene expression were observed in left-pawed and right-pawed dogs, but not in ambidextrous dogs.

Considering now the avian class, it has been reported that during hatching the embryos of many precocial bird species turn the body anticlockwise (see Rogers, 2006) as they break the eggshell, driven by the right side of the body as the right leg, head and neck cause rotation within the shell (Bekoff and Kauer, 1984; Casey, 2005). Subsequent to hatching, several precocial species of birds show a preference for using the right foot to initiate ground scratching while searching for food (Rogers and Workman, 1993). Given that the major force during anticlockwise full body turns made to escape from the eggshell during hatching is exerted by the right foot, this could be one explanation for the presence of right-footedness after hatching (but see Chapter 4 for an alternative explanation of the asymmetry of food searching behaviour – embryo orientation and the effects of light on the development of asymmetry in birds). It should be noted, however, that postural and positioning control could be much more demanding than the simple reflex-like activities associated with ground scratching. Tommasi and Vallortigara (1999) recorded foot use during ground scratching in domestic chicks searching for food under normal binocular vision and in animals with an eye temporarily occluded by an eye-patch. Binocular chicks showed a significant right foot bias, whereas monocular chicks tended to use the foot opposite to the eye in use, i.e. the right foot while seeing with the left eye and the left foot while seeing with the right eye. Data collected from monocular chicks thus suggest that the activated hemisphere (contralateral to the eye in use) is the one that takes control of posture, leaving to the other hemisphere reflex-like responses associated with ground scratching or body wiping. Footedness in birds might have arisen from the limb that is used to maintain postural and positional control, rather than from the limb that is used during motor activities such as ground scratching.

Asymmetrical use of the limbs is likely to be very widespread among vertebrates, even in taxonomic groups with little usage of these appendages. Bisazza et al. (1996) found a significant population preference for the use of the right forepaw in the common European toad Bufo bufo in a task in which the toads had to remove an elastic balloon or a strip of paper from the head. In another test, it was shown that the South American cane toad Bufo marinus used the right forepaw preferentially to control rolling to an upright position after the body had been turned over and submerged in water. Asymmetric use of the limbs also occurs for hindlimbs, which are not used in any wiping behaviour (see Malaschichev and Wassersug, 2004 for a review).

Some form of 'handedness' might have appeared even before the evolution of tetrapods. The blue gourami fish (Trichogaster trichopterus) uses its ventral fins as appendages serving gustatory and tactile functions with which it inspects nearby objects. When exposed to a sequence of novel objects, varying in shape and colour, these fish showed preferential use of the left fin during initial contact (Bisazza *et al.*, 2001).

## 1.3 Emotion and motivation

It is important to understand whether a general and common pattern of lateralization can be observed in all vertebrates or if lateralization has evolved many times under similar evolutionary pressures (see for a discussion on this issue Vallortigara, 2005, and papers cited therein). With respect to emotional and motivational aspects of behaviour, the overall evidence from different taxonomic groups seems to be clearly in favour of a common and shared pattern. Let us consider briefly some evidence for this.

Amphibians (e.g. toads), Australian lungfish (Lippolis *et al.*, 2002, 2009), mammals (e.g. dunnarts; Lippolis *et al.*, 2005) and birds (e.g. domestic chicks; Rogers, 2000) all appear to be more reactive to predators seen in their left, rather than right, visual hemifield (i.e. when their right hemisphere is attending to the predator stimulus; reviews in Rogers and Andrew, 2002; Vallortigara, 2006a; MacNeilage *et al.*, 2009; Rogers, 2010a). In rats, lesions to the right hemisphere elevate activity in the open field (Robinson, 1985), as a result of suppression of the fear response of freezing. The heightened anxiety that occurs in humans following damage to the right hemisphere may have similarity with increased activity responses shown by animals following the same lesion (Robinson and Downhill, 1995). Neurochemical changes have been observed in the right, but not the left, hemisphere associated with predator stress in rats and cats (Adamec *et al.*, 2003).

In contrast to the leftward bias in responses to predators, a rightward bias for feeding responses has been documented in a variety of species (from toads and fish to different species of birds; Rogers and Andrew, 2002). These rightward biases for prey catching and foraging responses are apparent when the prey or food has to be discriminated from similar targets; for example, toads show a right hemifield preference for directing tongue strikes at prey that has to be recognized precisely and 'handled' with care (e.g. crickets) but not for simplified prey models, such as a rectangular silhouette moving along its longitudinal axis (Robins and Rogers, 2004). Similarly, chicks show a rightward bias for pecking at grain (controlled by inputs from the right eye) that has to be discriminated against a distracting background (Figure 1.4) or uncovered by removing a lid from a bowl (Andrew *et al.*, 2000). These findings indicate that the left hemisphere, which primarily processes input from the right eye, controls responses that require considered discrimination between stimuli and/or manipulation of objects.

Figure 1.4 The pebble floor task, in which chicks discriminate between grains of food and pebbles scattered on the floor, revealed better performance by chicks using their right eye (left hemisphere). Photograph courtesy of Dr Cinzia Chiandetti, laboratory of Professor Giorgio Vallortigara.

Aggressive responses seem to be mainly controlled by the right side of the brain. During intraspecific agonistic encounters, animals direct more aggressive responses to conspecifics on their left side than they do to those on their right side (e.g. toads, Robins *et al.*, 1998; Robins and Rogers, 2004; lizards, Hews and Worthington, 2001; chicks, Vallortigara *et al.*, 2001; gelada baboons, Casperd and Dunbar, 1996; horses, Austin and Rogers, 2012). In chicks, intracranial treatment with glutamate (an excitatory neurotransmitter, which impairs neural development) revealed that expression of aggression is inhibited by the left hemisphere (glutamate treatment of the left hemisphere causes the release of aggressive responses; discussed in more detail in Chapter 4). A similar inhibition of the right hemisphere by the left hemisphere has been shown in humans: impairment of the left hemisphere leads to the expression of more intense emotions controlled by the right hemisphere (Nestor and Safer, 1990). In other species, the right hemisphere's control of aggressive responses has been shown by a left eye preference immediately prior to attacking the opponent (e.g. Austin and Rogers, 2012).

Studies of self-directed behaviour have revealed a larger contribution of the left hand, compared with manual actions involving inanimate objects. For example, a left-hand bias for face touching has been reported in orang-utans (*Pongo pygmaeus*) (Rogers and Kaplan, 1996) and gorillas (*Gorilla gorilla*) (Dimond and Harries, 1984), and in a non-primate marsupial mammal, a wallaby (Giljov *et al.*, 2012). A greater involvement of the left side of the body during socially arousing situations compared with non-emotive situations may be due to the right hemisphere's dominant role in perceiving emotion, with differential effects on primary cutaneous nerves receiving stimulation across the left and right hemispaces of the integument (e.g. Hopkins *et al.*, 2006). Behavioural asymmetries tied to emotive stimuli have also been reported for perception and production of facial expressions in both apes and humans. These studies are consistent with a right hemisphere dominance for emotive stimuli and reports of an earlier activation of the left side of the face (e.g. Borod *et al.*, 1986; Fernandez-Carriba *et al.*, 2002).

There is, however, an interesting alternative to the simple hypothesis that all emotions are controlled by the right hemisphere. The so-called 'valence hypothesis' maintains that the right hemisphere is involved in the processing of negative emotions, whereas the left hemisphere is involved in processing positive emotions (Davidson, 1995). The hypothesis is supported by clinical observations that patients with damage to the left hemisphere are more likely to manifest depressive symptoms than those with damage in equivalent areas of the right hemisphere (Goldstein, 1939/1963) and that a pathological condition of laughing is sometimes seen in patients with damage to the right hemisphere (Sackeim *et al.*, 1982). The valence hypothesis can also be associated with the motivational approach–withdrawal hypothesis, which looks at the emotional system from an evolutionary perspective. According to this hypothesis, organisms should behave differently according to whether positive or negative emotions are expressed, with an approaching or withdrawing reaction, respectively. In dogs, Quaranta *et al.* (2007) showed that stimuli that could be expected to elicit approach tendencies (such as a dog seeing its owner) are associated with higher amplitude of tail-wagging movements to the right side (left brain activation because of contralateral neural control of muscles producing tail movements) and stimuli that could be expected to elicit withdrawal tendencies (such as seeing a dominant unfamiliar dog) are associated with higher amplitude of tail-wagging movements to the left side (right brain activation) (see Figure 1.5).

It has been suggested that the right hemisphere of the human brain is related to sympathetic arousal and the left hemisphere to parasympathetic calmness (Craig, 2005). This idea would be consistent with the proposal that the right hemisphere responds to the unexpected (novel) stimuli and the left hemisphere to more routine (familiar and less arousing) stimuli (MacNeilage *et al.*, 2009). It is interesting to note in this regard an important anatomical asymmetry documented recently in the brains of both apes and humans, concerning the so-called von Economo neurons (VENs) (see von Economo and Koskinas, 1925). The von Economo neurons are large bipolar neurons located in frontoinsular (FI) and

**Left hemisphere**

Viewing owner

Approach

**Right hemisphere**

Viewing dominant dog

Withdrawal

Figure 1.5 Tail wagging in dogs. Measurements of amplitude of tail wagging in dogs looking at different types of emotional stimuli have shown that a larger excursion of tail wagging to the dog's right side occurs when the dog views its owner and to the left side when it views a dominant dog (Quaranta *et al.*, 2007). Wagging to the right is associated with approach (left hemisphere). Wagging to the left (right hemisphere) indicates emotional fear and likelihood of either withdrawing or adopting defensive threat behaviour.

anterior cingulate cortex (ACC), and they seem to be present in great apes and humans. It is believed that VENs may allow rapid communication across relatively large brains (in fact, they have also been observed in elephants and whales and have been implicated in many cognitive abilities associated with social behaviour and decision-making; Allman *et al.*, 2011). In particular, the anterior insula and cingulate cortex are involved in the recognition of error and the initiation of adaptive responses to error and negative feedback. Interestingly, VENs are more numerous in the right hemisphere in both humans and great apes (Allman *et al.*, 2011; no evidence is currently available for possible asymmetries in elephants and whales). Allman *et al.* (2011) suggested that VEN asymmetry may be related to asymmetry in the autonomic nervous system in which the right hemisphere is preferentially involved in sympathetic activation, as would result from negative feedback and subsequent error correcting behaviour, and the left hemisphere is preferentially involved in parasympathetic activity associated with reduced tension or calming responses (Craig, 2005). Following on from this reasoning, Allman *et al.* (2011) suggested that there may be more VENs on the right side, because the responses to negative feedback require more complex and more urgent behavioural responses than do situations that are calming and involve reduced tension.

## 1.4 Cognition

Cognition too has revealed a basically similar pattern of lateralization in vertebrates. This has been investigated mainly in three basic domains: categorization processes, social cognition and spatial cognition.

Lateralization of visual categorization in birds has been widely documented using the pebble floor task, mentioned above. Chicks using their right eye performed the task better than those using their left eye; similar results have been obtained in pigeons (Güntürkün and Kesh, 1987), zebra finches (Alonso, 1988) and quails (Valenti *et al.*, 2003). Moreover, using conditioning procedures it has been shown that the right eye/left hemisphere is dominant in tasks involving colour, shape and object discriminations (e.g. Diekamp *et al.*, 1999; Vallortigara *et al.*, 2004). In fact, the two hemispheres provide different contributions to concept formation. For instance, Yamazaki *et al.* (2007) trained pigeons to categorize pictures of humans and then tested them (binocularly or monocularly) for transfer to novel exemplars. The authors examined (i) whether the two hemispheres relied on memorized features or on a conceptual strategy, using stimuli composed of new combinations of familiar and novel humans and backgrounds; (ii) whether the hemispheres processed global or local information, using pictures with different levels of scrambling; and (iii) whether they attended to configuration, using distorted human figures. The results suggested that the left hemisphere employs a category strategy and attends to local features, whereas the right hemisphere uses an exemplar strategy and attends to configuration.

Cognition and navigation in space provide further evidence for basic dichotomies in hemispheric function. Vallortigara *et al.* (2004) trained chicks binocularly in a rectangular enclosure with panels at the corners providing cues for food localization in one of the corners. When tested after removal of the panels, chicks using their left eye (with a patch over their right eye), but not chicks using their right eye, reoriented using the information provided by the geometry of the cage (i.e. combining metric properties such as length of the walls, short *vs.* long, with sense direction, left *vs.* right). In contrast, when tested after removal of geometric information (i.e. in a square-shaped cage), both right- and left-eyed chicks reoriented using the non-geometrical information provided by the panels. When trained binocularly with only geometric information (in a rectangular enclosure without any distinctive panels at the corners) and tested for recall, chicks using their left eye reoriented better than those using their right eye. Finally, when geometric and non-geometric cues provided contradictory information (because of an affine transformation on the spatial distribution of panels), left-eye-using chicks showed more reliance on geometric cues, whereas right-eye-using chicks showed more reliance on non-geometric cues. These results suggest the existence of separate mechanisms for dealing with spatial reorientation, the right hemisphere attending to the large-scale geometry of the environment and both hemispheres attending to local, non-geometric cues when available in isolation, but with a predominance of the left hemisphere when competition between geometric and non-geometric information occurs.

Hemispheric differences in spatial cognition similar to those observed in chicks have been reported in rats (LaMendola and Bever, 1997). Data suggest dissociations along similar lines in humans. For instance, right hippocampal activation has been documented in taxi drivers asked to mentally navigate the streets of London (Maguire *et al.*, 2006). Using the rectangular room task mentioned above, Pizzamiglio *et al.* (1998) showed that patients with right brain damage and left hemineglect (i.e. selective inability to attend to the left hemispace) are impaired in spatial reorienting.

Evidence for different mechanisms dealing with spatial information in the right and left hemispheres also comes from place-finding tasks. Chicks can be trained to localize the central position in an enclosure in the absence of any external cues and are able to generalize among enclosures of different sizes (Tommasi *et al.*, 1997). However, when, after training in a square-shaped enclosure, chicks were tested in an enclosure of the same shape but of a larger size, their search in the larger enclosure was localized in two regions: in the actual centre of the test enclosure and at a distance from the walls that was equal to the distance from the walls to the centre in the training (smaller) enclosure (Tommasi and Vallortigara, 2001). Encoding of the goal location, therefore, seems to occur in terms of absolute distance and direction to the walls and in terms of ratios of distances (whatever their absolute values) from the walls. Tests carried out under monocular viewing (after binocular training) revealed a striking asymmetry: the left hemisphere attended to absolute distances, whereas the right hemisphere attended to relative distances (Tommasi and Vallortigara, 2001). This suggests that both hemispheres are involved in encoding of spatial cognition, though with basically different strategies.

Lateralization of spatial function in birds seems to rely, as it does in humans (e.g. Maguire *et al.*, 2006) and rodents (e.g. Samara *et al.*, 2011; Shinohara *et al.*, 2012), on asymmetry of hippocampal function. This suggests basic homology because the hippocampal formation is an ancient structure inherited in birds and mammals from reptilian ancestors (and there is evidence of the presence of it in modern teleost fish as well; see Rodriguez *et al.*, 2002). Tommasi *et al.* (2003) trained chicks with the hippocampus lesioned either bilaterally or unilaterally to search for food hidden beneath sawdust in the centre of a large enclosure, the correct position of food being indicated by a local landmark. At test, the land-mark was removed or displaced at a distance from its original position. Results showed that sham-operated chicks (those with surgery but without a lesion) and chicks with a lesion to the left hippocampus searched by scratching the ground/sawdust in the centre, relying on large-scale spatial information provided by the enclosure, whereas chicks with a lesion of either the right hippocampus or both hippocampi were completely disoriented (landmark removed) or searched close to the landmark shifted from the centre (landmark displaced). These results indicate that encoding of spatial features of an enclosure occurs in the right hippocampus even when local information provided by a landmark would suffice to find the goal; encoding based on local information (landmark), in contrast, seems to occur in regions of the brain other than the hippocampus.

Recordings of the activity of single neurons in the left and right side of the brain may provide some insights on the mechanisms of lateralized spatial cognition. Folta *et al.* (2004) recorded the activities of neurons in the left or the right nucleus rotundus in the pigeon while stimulating the ipsilateral or the contralateral eye (Figure 1.6). The authors were able to distinguish between

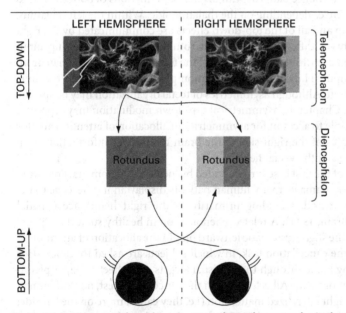

Figure 1.6 Bottom-up and top-down inputs to the nucleus rotundus in the pigeon. The nucleus rotundus is a diencephalic structure. Asymmetries in top-down projections, from the telencephalon to the diencephalon, make the left nucleus rotundus mainly represent the afferents from the right eye, while the right nucleus rotundus integrates input from both eyes. The figure is based on the single cell recording data of Folta *et al.* (2004).

left–right differences that emerged bottom-up from the retino-tecto-rotundal system and those that were derived top-down from the forebrain. Left–right differences within the bottom-up system were due to variations in the latency and the tonic spike duration of rotundal neurons after stimulation of the contralateral eye. Visual signals arrived on average 18% faster in the right thalamus, but cellular activation lasted 27% longer in the left rotundus. The authors suggested that these lateralized effects may underlie the fact that pigeons are faster with the left eye (right hemisphere) in simple visual reaction paradigms, but are superior with the right eye (left hemisphere) in pattern learning and discrimination. While the asymmetries within the bottom-up system were a matter of degree, those of the top-down cells displayed an all-or-none organization: all thalamic cells that were activated by descending forebrain systems were under control of the left hemisphere. Thus, although visual input reaches both hemispheres, the modulation of the diencephalic relay of the tectofugal system seems to be under the executive control of the left hemisphere only.

An intriguing consequence of this organization is that if descending forebrain signals arrive within the rotundus only from the left hemisphere, they should produce response patterns with diverse combinations of bottom-up and top-down influences depending on the thalamic side. Within the left thalamus, most bottom-up and all of the top-down effects are communicated by the right eye system. This is different for the right rotundus, where bottom-up input derives from left eye stimulation while all of the top-down effects originate from the right eye input. Thus, bilateral integration predominates at right thalamic level but not at left. Although asymmetries of tectal organization may be specific to pigeons (see Chapter 4), asymmetrical top-down modulation may represent a general principle to account for asymmetries in allocation of attention and for the better abilities of the right side of the brain to consider information from widespread areas of the visual field.

An example of the first case is represented by 'neglect' phenomena. Following right hemisphere damage, many human patients display an indifference to the left side of the world, attending primarily to the right hemispace ('spatial hemineglect'; Brain, 1941). A related phenomenon in healthy subjects is 'pseudoneglect', i.e. the slight systematic leftward bias in the allocation of attention in tasks, such as the cancellation task, in which subjects are asked to 'cancel' (for instance putting a line through them) visual targets on a sheet of paper placed midline in front of them (Albert, 1973). In the cancellation test, normal human subjects show right lateralized inattention (i.e. they cancel more on the left side; Vingiano, 1991). Diekamp et al. (2005) devised an adapted version of this task and administered it to domestic chicks and pigeons. Birds were given free choice to orient towards and peck at grains that were spread evenly over an area in front of them. The animals could move their head freely while their body was restrained and aligned centrally in front of the search area (Figure 1.7).

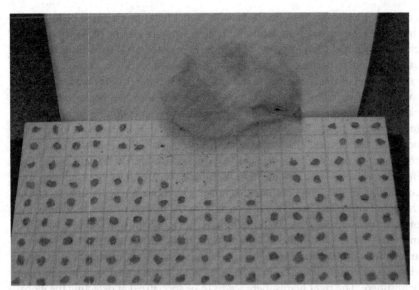

Figure 1.7 A 'cancellation task'. The chick's head is poking through a hole in a wall. At the start of the test, one grain of chick mash is in each square on the floor. The chick can move its head and peck wherever it likes. The midline is on row nine of the array. Note the leftward bias in grains consumed: twice as many have been taken on the left as on the right side. Photograph courtesy of Dr Cinzia Chiandetti, laboratory of Professor Giorgio Vallortigara.

Both chicks and pigeons showed a leftward bias. In a similar vein, adult nutcrackers (*Nucifraga columbiana*) and newborn domestic chicks show a leftward bias when required to locate an object in a series of identical ones on the basis of its ordinal position (Rugani *et al.*, 2010, 2011). In these studies birds were trained to peck at either the fourth or sixth element in a series of 16 identical and aligned positions. These were placed in front of the bird, sagittally with respect to its starting position. When, during testing, the series was rotated by 90° lying frontoparallel to the bird's starting position, both species showed a bias for identifying selectively the correct position from the left but not from the right end. A similar though stimulus-specific hemineglect has been reported in toads (Vallortigara *et al.*, 1998), which are more likely to attack prey to their right side (and ignore them to their left side) and to attack conspecifics on their left side (and ignore conspecifics on their right side).

The electrophysiological data of Folta *et al.* (2004) also agree with behavioural evidence from studies of the abilities of young chicks to mentally complete partially occluded objects, a cognitive feat that requires the integration of information from widespread areas of the visual field (reviewed by Vallortigara, 2004, 2006b). Data collected both in young domestic chicks and adult hens suggest that this species is capable of perceiving as a whole objects that are partly concealed by occluders ('amodal completion'). Regolin et al. (2004) investigated hemispheric differences in amodal completion by testing separate groups of newly hatched chicks which had been imprinted binocularly on a red cardboard square, on the same red square partly occluded by a superimposed black bar, or on an amputated version of the red square (i.e. a red square lacking the very same area that would be occupied by the superimposed bar). At test, in monocular conditions, each chick was presented with a free choice between a complete and an amputated square. In the crucial condition following imprinting with the partly occluded red square, chicks tested with only their left eye in use chose the complete square (like binocular chicks); chicks with only their right eye in use, in contrast, tended to choose the amputated square, thus not showing any completion of the partly occluded stimulus. These results suggest that in chicks the right hemisphere is mainly responsible for the process of amodal completion of partly occluded objects. Even in humans the right hemisphere seems to play a more important part in amodal completion (Corballis et al., 1999). In order to amodally complete an object, the spatial relationships between the parts of a visual scene must be taken into account, a task performed well by the right hemisphere. The right hemisphere seems to be specialized at detecting the global structure of visual objects, whereas the left hemisphere seems to be more inclined to detect local features.

With regard to lateralization in response to social stimuli, the record of comparative data is quite impressive (reviewed by Vallortigara and Andrew, 1994a; Rosa Salva et al., 2012). In humans, right hemisphere dominance in face perception is well attested in the neuropsychological literature (Bradshaw and Nettleton, 1982; and for studies on functional images see Kanwisher et al., 1996). Similar evidence has been reported in studies with non-human primates. Dominance of the right hemisphere in face recognition has been found by testing split-brain monkeys (Hamilton and Vermeire, 1988), and fMRI studies have demonstrated a more pronounced activation of the right superior temporal sulcus (Pinsk et al., 2005). Like humans, apes exhibit a left visual field (LVF) advantage for the recognition of facial emotional expressions (Morris and Hopkins, 1993; see also Fernandez-Carriba et al., 2002). This right hemispheric dominance for the processing of emotional expressions may be more pronounced for negative than for positive emotions (e.g. Parr and Hopkins, 2000).

A prominent phenomenon associated with right hemispheric dominance in face perception is the left gaze bias (LGB), a tendency to spend a higher proportion of looking time exploring the left side of a centrally presented face or to direct the first fixation towards the left side of the face. Guo et al. (2009) reported a LGB in both humans and rhesus monkeys (Macaca mulatta).The LGB was also found in pet dogs (Canis familiaris) for human faces (Guo et al., 2009).

In sheep, behavioural studies employing mirrored hemifaces and chimeric faces (i.e. faces composed of two right or two left hemifaces) as stimuli demonstrated a left visual field (LVF) advantage for individual recognition of faces of familiar conspecifics but not for human faces (Peirce *et al.*, 2000, 2001; Kendrick, 2006). (The lack of LVF advantage for the processing of human faces in sheep could be due to lack of, or insufficient, exposure to human faces early after birth.) It is interesting to note that this category (faces of familiar conspecifics) is exactly the category of face stimuli that sheep can recognize thanks to the configuration of inner face features, in line with that usually found in human beings for the processing of own-species faces. The right hemisphere advantage found in sheep can thus be explained by a general superiority of the left eye system for configurational processing of visual stimuli (as hypothesized also for human beings; Levy *et al.*, 1972). Yamazaki *et al.* (2007) found evidence of an analogous configural processing style of the right hemisphere, in pigeons trained to respond to pictures of human beings. In this study pigeons were reinforced for discriminating between pictures containing and pictures not containing human figures. After that they underwent a generalization test on novel exemplars of the same categories. Pigeons with their left eye (right hemisphere) in use were severely disturbed by manipulations altering the overall stimulus configuration (such as the use of scrambled stimuli), whereas subjects with the left hemisphere in use tended to base their choice on local category-defining cues.

Expression of activity-dependent genes (c-fos and zif/268) in brain regions specialized for processing faces (temporal and medial frontal cortices and basolateral amygdala) also revealed greater activation of the right hemisphere in sheep undergoing a discrimination task of upright faces, whereas the effect was absent for inverted faces (Broad *et al.*, 2000). In this case, the task required discrimination of pictures of human versus sheep faces (the reward for a choice was gaining access to the photographed individual). Right hemisphere activation was observed also in individuals that did not demonstrate a preference for sheep faces in terms of the percentage of choices for the two kinds of faces (however, latency data revealed an advantage of sheep faces in all subjects, including this subgroup). Similar results were also obtained in a subsequent study, in which socially isolated sheep were simply exposed to images of a conspecific face belonging to their own breed (Da Costa *et al.*, 2004). Activation of the right hemisphere was elicited by faces of same-breed conspecifics, but not by exposure to goat faces. This effect appeared to be associated both with face processing and with emotional responses determined by the exposure to conspecific faces (having a calming effect on distressed animals).

Neurons selective for faces are present in the temporal cortex of sheep, similar to the location usually found in monkeys (Kendrick and Baldwin, 1987; Kendrick *et al.*, 1995, 1996, 2001). Studies on sheep provided evidence of a temporal advantage of face cells in the right hemisphere over left hemisphere ones (Peirce and Kendrick, 2002), in line with electrophysiological results in humans (Seeck *et al.*, 1997). In particular, cells that respond selectively to one type of face or to the face of one individual show reduced response latency when processed in the right hemisphere. This temporal advantage is limited to 'subtle' discriminations between different faces, whereas the two hemispheres are equally fast in simply detecting the presence of a face in the visual field. Moreover, a certain number of neurons in the left hemisphere actually respond after the time necessary for sheep to identify a face is elapsed. It seems reasonable that these slower cells are not contributing critically to visual recognition, but rather are involved in later processing contingent on recognition (e.g. processing of behavioural responses appropriate to the recognized face).

Research on domestic chicks and other species has provided consistent evidence of behavioural lateralization of social recognition. Using free choice tests and social pecking tests in animals with vision confined to only the right or the left eye, it has been shown that chicks are better at discriminating companions from strangers when using their left eye (right hemisphere, Vallortigara and Andrew, 1991; Vallortigara, 1992; Vallortigara et al., 2001; Deng and Rogers, 2002b). The results were also confirmed in adult hens (McKenzie et al., 1998; and see Zucca and Sovrano, 2008 for evidence in adult quails). Furthermore, Town (2011) recently reported that IMM neurons in chicks respond differentially to videos of conspecifics based on past experience and that the disparity between responses to familiar and unfamiliar conspecifics is greater in the right than left IMM.

Further evidence for a role of the right hemisphere in social recognition also comes from studies on transitive inference (Daisley et al., 2011) and social learning in chicks (Rosa Salva et al., 2009). A case in point is the ability of domestic chicks to perform transitive inference, which is associated with social group formation and dominance hierarchies. Daisley et al. (2010) trained chicks to discriminate stimulus pairs (A vs. B, B vs. C, C vs. D, D vs. E), in order to build a hierarchy (A > B > C > D > E). Chicks were subsequently tested on a stimulus pair never seen together before (BD). Chicks using their left eye only (right hemisphere) during test showed better performance than did right-eye-only (left hemisphere) chicks on the BD task.

A simple method to reveal the presence of behavioural asymmetries in social cognition, which has been used extensively in species such as fish, amphibians and reptiles, involves recording the frequency of social responses (e.g. aggressive or sexual behaviours) occurring towards conspecifics when they appear on the left versus on the right side of the subject (i.e. fall within the left or the right visual field). In most cases, aggressive and sexual displays are performed towards conspecifics and imply recognition of the other organism as belonging to the same species as the subject (sometimes also recognition of the gender or individual identity of the other organism is required). With this procedure it has been demonstrated that, in female striped plateau lizards (Sceloporus virgatus, males were not tested), aggressive displays towards courting males are more frequent if the male appears on the left side of the subject, or in its binocular visual field, than if the male appears on its right side (Hews et al., 2004). Similarly to that described above for chicks, performance of lizards under the guidance of their binocular visual field resembled that observed under the guidance of the LVF, suggesting that the right hemisphere could determine performance in binocular vision conditions. Moreover, female lizards also show a bias for orienting to view conspecifics with their LVF before performing a charge (a strongly aggressive act) (Hews et al., 2004). This general pattern of behaviour was evident both in controlled laboratory conditions and in unmanipulated natural encounters.

Similar data to those obtained in females by Hews and collaborators (2004) are also available for males in another species of lizards (Urosaurus ornatus), which show more aggression towards other males when in the LVF (Hews and Worthington, 2001; see also Deckel, 1995 for evidence on male green anoles, Anolis carolinensis). Moreover, studies on lizards suggest that individuals in different motivational states may use brain lateralization to alter the nature of social encounters. That is to say, lizards that are likely to be aggressive (e.g. because they are bigger than their opponent; see Hews and Worthington, 2001) could decide to use the LVF for their aggressive display, turning their heads so as to view their rival on their left side. This could in turn modify their own behavioural response (determining increased aggressiveness) and signal to the opponent that they are unlikely to give up the fight.

Research conducted on lizards also revealed that an asymmetry in endogenous serotonin (5-HT) levels (possibly in the raphe nucleus) is likely to contribute to the behavioural asymmetries described in the previous paragraph. Consistent with that is the fact that quipazine, a 5-HT2 agonist, decreases selectively right hemisphere-stimulated aggression (Deckel and Fuqua, 1998). Other evidence consistent with this idea comes from the fact that mild stressors and experimentally induced alcohol withdrawal can decrease the leftward bias in aggression (Deckel, 1997, 1998; Deckel et al., 1998). Both these manipulations might act on the asymmetry in endogenous serotonin. In line with that, evidence available in the literature suggests greater sensitivity of the right hemisphere than of the left hemisphere to alcohol administration (e.g. Erwin and Linnoila, 1981): there is impairment of the right hemisphere's cognitive functions other than that of social recognition (e.g. spatial abilities in rats, Blanchard et al., 1987).

A left eye bias during aggressive or courtship behaviour has been demonstrated in toads (Bufo marinus, Robins et al., 1998; Bufo bufo, Vallortigara et al., 1998) and also in avian species (e.g. Rogers et al., 1985; Ventolini et al., 2005) and non-human primates (Theropitecus gelada, Macaca mulatta, Papio cynocephalus, Casperd and Dunbar, 1996; Drews, 1996).

Results in line with the general evidence of a left visual hemifield bias for guiding social responses and social recognition have been obtained by testing fish with slightly different procedures that investigate side preferences without involving clear aggressive or sexual behaviours. For example, eight species of teleost fish (Sovrano et al., 1999), as well as the tadpoles of five anuran species

(Bisazza *et al.*, 2002), show a left eye preference for inspection of their mirror image, perceived as an unfamiliar conspecific (see also De Santi *et al.*, 2001; Sovrano *et al.*, 2001; Sovrano and Andrew, 2006). The mirror test exploited in these studies, initially used to prove that fish (*Gambusia holbrooki*) were more prone to show predator inspection when they shared with a social partner the risk of being preyed on, has also revealed that predator inspection is more frequent when the social companion (either a mirror image or a video recorded stimulus) is visible on the left side (Bisazza *et al.*, 1999).

Also in this case, as previously discussed for face perception studies, the role of experience with a given stimulus (or stimulus category) is important in determining the preferential involvement of the right hemisphere. It is worth mentioning that, in fish, the same left eye bias demonstrated in the mirror test is evident also for the inspection of familiar abstract patterns, but not for unfamiliar ones (Sovrano, 2004). However, when social stimuli are employed, such as in the mirror test, fish display a preferential left eye use regardless of whether they have been directly familiarized with their own mirror reflection: a period of visual experience either with other conspecifics or with their own mirror reflection is sufficient to determine this effect (Sovrano, 2004). Thus, for fish, visual familiarity with the category to which the stimulus belongs to (e.g. same-sex conspecifics) is enough to determine a right hemisphere involvement. This would be remarkably similar to what has been hypothesized for face perception in primates and sheep. In fact, such species seem to develop a special expertise for the type of face more extensively experienced during (early) life, with preferential involvement of the right hemisphere for this kind of face only. Nevertheless, it is also possible to hypothesize that fishes would use their left eye when looking at familiar stimuli or at stimuli presenting only a limited degree of novelty: unknown conspecifics could be characterized exactly by this moderate degree of novelty. In general, results obtained with the mirror test in fishes are in agreement with those reviewed previously for other species in that the right hemisphere dominance (for monitoring conspecifics according to an affiliative motivation) seems to be linked to the specialization of this hemisphere for the evaluation of stimuli's familiarity.

Evidence of side biases in naturalistic encounters that are neither sexual nor aggressive is also present in avian species and in mammals. One notable example is that of pigeons: when homing in flocks, pigeons tend to fly to the right of their favourite flight partners, monitoring them with their left eye (i.e. the longer a bird had spent behind another individual, the more likely it was to be flying to the partner's right side) (Nagy *et al.*, 2010). Thanks to sophisticated analyses of pigeons' flight trajectories, Nagy and collaborators concluded that when pigeons elaborate visual cues from a given conspecific predominantly with their left eye, they are more responsive to its movements, which influence the pigeon's own flight trajectory.

Similarly, gelada baboons (*Theropithecus gelada*) prefer to process social visual cues from conspecifics with their right hemisphere even during nonaggressive approaches (Casperd and Dunbar, 1996). In sea mammals, a bias to monitor social signals from conspecifics with the left eye has been observed during mother–offspring interactions: in the natural environment, calves of the wild beluga whale (*Delphinapterus leucas*) swim keeping their mother on the left side (Karenina *et al.*, 2010; see also Sakai *et al.*, 2006 for similar evidence of more pronounced social rubbing behaviour with the left than with the right flipper, associated with a preferential left eye use, in bottlenose dolphins *Tursiops aduncus*).

Overall, the data on lateralization on social cognition in animals suggest that left visual hemifield bias is probably part of a more general specialization of the right hemisphere to establish identity, i.e. that an apparently familiar stimulus is indeed identical to one experienced previously. Preferential use of the monocular field of the right eye, in contrast, seems be associated with visual control of response when a rapid decision should be taken. For instance, in zebrafish, right eye use goes with visual control of response (e.g. when zebrafish are about to approach and bite a target). Right eye use when viewing a predator (De Santi *et al.*, 2001) can be explained similarly: again the fish is ready to perform a response (escape) whilst viewing the object that may evoke the response.

Although we have concentrated this brief review on lateralization of visual perception and processing, it is important to say that similar asymmetries have been documented in other sensory modalities in animals (e.g. olfaction, Vallortigara and Andrew, 1994b; audition, Andrew and Watkins, 2002). Also, there are other specialized forms of behaviour, such as homing in pigeons, that have revealed striking brain lateralization but are not discussed here (see for a recent review Pecchia *et al.*, 2012).

To sum up, a common pattern of lateralization is apparent among vertebrate species. This is roughly summarized in Figure 1.8 but it will be discussed again and refined in Chapters 2 and 5. Briefly, the left hemisphere is specialized to attend to similarities or invariances between stimuli, in order to allocate stimuli into categories following rules established through experience or biological predispositions (see also Vallortigara *et al.*, 2008). The left hemisphere shows focused attention, in particular on local features of the environment, so that the animal is not easily distracted by extraneous stimuli. In fact, one can say that the left hemisphere performs routine functions and establishes patterns of behaviour in non-stressful conditions, i.e. under parasympathetic quietude. The right hemisphere, on the other hand, attends to novel stimuli (variance). It notices unique and small differences between stimuli and, as an aspect of this specialization, it is easily distracted from the task being performed. It is the hemisphere that takes charge of behaviour in emergency situations, expressing heightened fear and aggression. It is also the hemisphere that expresses intense emotions, particularly negative ones, under sympathetic arousal. Along with this, the right hemisphere shows diffuse attention making it specialized to attend to the global rather than the local properties of stimuli, as shown both

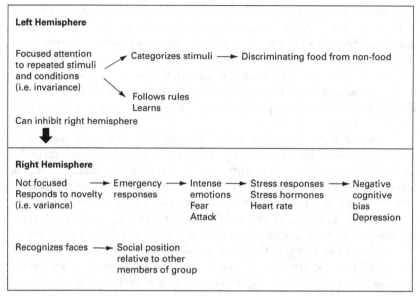

Figure 1.8 Summary of left–right differences. The main differences in function between the left and right hemispheres in vertebrate species have been summarized. From Rogers (2011a).

in spatial and social cognition. In particular the right hemisphere is specialized in individual recognition (for instance face recognition), very likely as part of a general specialization to attend to social relationships.

## 1.5 Lateralization in invertebrates

Recently, evidence that brain and behavioural asymmetries are not limited to vertebrates has begun to emerge. Honeybees (*Apis mellifera*), in particular, have provided clear evidence for functional lateralization. Letzkus *et al.* (2006) conditioned bees to extend their proboscis in anticipation of a food reward when they perceived a particular odour. Some bees had their left antenna covered with a silicone compound, which prevents detection of odour, others had their right antenna covered, and those in a third group constituted a control in which both antennae were uncovered. Bees with the right antenna covered learned less well than the bees with their left antenna covered and bees with both antenna uncovered.

Electron microscope analyses revealed that there were a slightly higher number of olfactory sensilla in the right than in the left antenna (Frasnelli *et al.*, 2010). However, it seems unlikely that this can account for the functional asymmetry. Rogers and Vallortigara (2008) extended the work of Letzkus *et al.* (2006) by investigating whether lateralization could be found in recall of olfactory memory at various times after the bees had been trained using both antennae (i.e. without forcing them to learn with one or the other antenna). After proboscis extension response (PER) training with both antennae in use, bees were tested for recall 1–2 and 23–24 hours later and with the left or right antenna coated with the silicone compound. At 1–2 hours after training, bees showed excellent recall when tested using their right antenna, but poor or no recall when tested using their left antenna. By contrast, 23–24 hours after training, recall was good when the left antenna was in use but not when the right antenna was in use, demonstrating that long-term memory is accessed mainly via the left antenna. Hence, retrieval of olfactory learning is a time-dependent process and involves lateralized neural circuits. Rogers and Vallortigara (2008) also checked whether the laterality was manifested as side biases to odours presented to the left or right side of the bee without any covering of the antennae. Bees were trained using both antennae, and the recall at several intervals (1, 3, 6 or 23 hours) after training was tested using lateral presentation of the two stimuli and no coating of the antennae (see Figure 1.9).

There were more correct PERs to odours presented on the right than on the left side at 1 hour after training. No significant left–right difference occurred at 3 hours after training. At both 6 and 23 hours after training the correct responses were higher on the left side than on the right side. The study by Rogers and Vallortigara (2008) thus demonstrates that the asymmetry is more complex than a difference in learning ability of the right and left antennae and

Figure 1.9 Lateralization in bees. Procedure used for testing bees with lateral presentation of an odour stimulus without any covering of antennae. The odour (e.g. lemon or vanilla) is dissolved in the droplet held to the left or right side of the bee. The response recorded is extension of the proboscis, the organ through which the bee ingests nectar or sugar solution. From Rogers and Vallortigara (2008).

that the difference in number of olfactory sensilla is unlikely to explain entirely the behavioural laterality. Up to now, however, search for anatomical correlates of the asymmetry in higher centres of the bee brain has not revealed clear anatomical asymmetries (Rigosi *et al.*, 2011; Haase *et al.*, 2011a, 2011b).

A shift of recall access from one to the other side of the brain has been noted previously in birds (Cipolla-Neto *et al.*, 1982; Clayton, 1993; Andrew, 1999). This is interesting because it suggests that lateralized events in memory formation may be similar in bees and vertebrate species. In bees, a shift of memory access from the right to the left side of the brain would allow the right antenna to learn about new odours without interference from odour memories in long-term stores. In fact, since bees visit different flowers at different times of the day, as nectar becomes available, the formation of different odour associations during the course of the day would be required, and this is a process that might be aided if recall of earlier odour memories is avoided on the side of the brain undergoing new learning.

Kells and Goulson (2001) reported that bumblebees *Bombus* spp. (Hymenoptera, Apidae, Apinae, Bombini) show preferred directions of circling as they visit florets arranged in circles around a vertical inflorescence (population-level lateralization is present). The biased circling might well have something to do with lateralization of antennal responsiveness to odours or lateralized learning and memory recall. In fact, a lateralization of PER learning similar to that of honeybees has been reported in bumblebees (Anfora *et al.*, 2011).

Further evidence of functional lateralization in invertebrates has been obtained in several sensory modalities and in motor control. Fruitflies (*Drosophila melanogaster*) showed a consistent asymmetry in antenna-mediated flight control: sensory signals from the left antenna contribute more to odour tracking than do signals from the right antenna (Duistermars *et al.*, 2009). Letzkus *et al.* (2007) used a PER paradigm to investigate visual learning of bees using their left or right eye. Bees seemed to use primarily their right eye for learning to associate a visual stimulus with a food reward. Ants (*Formicidae*) and spiders (*Araneae*) have been shown to possess behavioural left–right asymmetries (Heuts *et al.*, 2003). A significant majority of spiders in the field with mainly left leg lesions was observed, and also less severe leg lesions caused by catching them were significantly biased to the left. Appendage severance was significantly more frequent on the left than on the right side. Evidence of lateral biases in ants has been reported too (Heuts *et al.*, 2003). Twelve ant species of *Lasius niger* kept mainly to the right on their foraging 'streets', whereas there was only one species that kept to the left. On streets in trees, 49 *Lasius niger* colonies kept to the right versus 26 to the left. In this ant species a significant majority of couples in the laboratory had the left side of their bodies exposed to their partners when resting. This identical left body side exposure when resting and foraging in streets also correctly predicted that lone foraging *L. niger* would significantly more often turn to the right than to the left (the ratio was 14 to 2).

A behavioural asymmetry in mating behaviour, due to an anatomical asymmetry dependent on a maternal effect gene, has been observed in the pond snails *Lymnaea stagnalis* (Asami *et al.*, 2008; Davison *et al.*, 2009). The pond snail *Lymnaea stagnalis* is a self-fertile hermaphrodite; in any single mating an individual takes the male role or the female role. Chirality in snails is determined by the single locus of maternal effect (Boycott and Diver, 1923). This means that the phenotype of an individual is dependent upon the genotype of their mother. Asami *et al.* (2008) used crossing experiments to demonstrate that the primary asymmetry of *L. stagnalis* is determined by the maternal genotype at a single nuclear locus where the dextral allele is dominant to the sinistral allele. Dextral is dominant in *Lymnaea* (by convention, D = dextral allele; S = sinistral allele). The dextral and sinistral stocks are genetically DD or SS, respectively. On mating virgin sinistral and dextral snails, offspring that are genetically dextral (genotype = DS) but with a shell coil that is either sinistral (sinistral mother) or dextral (dextral mother) can be produced (F1 generation). By allowing the sinistral F1 mother to self-fertilize, offspring were produced that have a dextral coil, but are genetically DD, DS or SS (F2 generation). Dextral SS individuals were identified by virtue of their producing sinistral young. Davison *et al.* (2009) investigated the occurrence and the inheritance of a potential laterality trait in the pond snail and tried to understand whether it is associated with both body chirality and nervous system asymmetry. They found that all dextral snails circled in a counterclockwise manner, no matter whether they were paired with another dextral or a sinistral snail. Similarly, all the sinistral snails, both those paired with dextral and those paired with sinistral, circled in a clockwise manner.

Chirality in mating behaviour is matched by an asymmetry in the brain. *Lymnaea stagnalis* has a ring of nine ganglia that form a central nervous system around the oesophagus, with two more distant buccal ganglia on the buccal mass. In all dextral individuals, the right parietal ganglion was fused with the visceral ganglion, so that the left visceral ganglion was unpaired. By contrast, in all sinistral individuals, the reverse was observed; the left parietal ganglion was formed by fusion with a visceral ganglion. The central nervous system in sinistral pond snails, therefore, has an asymmetry that is reversed relative to that of dextral snails. Since the coil of the shell is determined by the maternal chirality genotype and the asymmetry of the behaviour is in accordance with this, it is likely that the same genetic locus, or a closely linked gene, determines the behaviour. These findings suggest that the lateralized behaviour of the snails is established early in development and is a direct consequence of the asymmetry of the body.

Evidence of a population-level lateralized behaviour has been found in the giant water bug *Belostoma flumineum* Say (Heteroptera: Belostomatidae; Kight *et al.*, 2008). The giant water bugs are large aquatic insects, predators of other aquatic invertebrates and small fishes. Bugs were trained to swim left or right in a T-maze and a significant preference to turn left, even when not reinforced, was observed. To control for environmental cues that might bias the turning direction of water bugs in the maze, the authors ran two separate experiments on independent groups of 20 water bugs. Both experiments were identical with the exception that, after the first group of 20 water bugs was tested, the maze apparatus was rotated 180° in the laboratory room, thus reversing the polarity of all directional environmental cues such as lighting or electromagnetic fields. Again the same left-turn tendency was observed. Hence, the explanation of the presence of this bias could be the existence of asymmetries in the nervous system or asymmetric exoskeletal morphology (i.e. leg length) that could cause biased swimming behaviour.

An asymmetry in T-maze behaviour has been reported in the cuttlefish (*Sepia officinalis*) trained to learn how to enter a dark and sandy compartment at the end of one arm of the maze (Alves *et al.*, 2007). The study revealed that 11 out of the 15 cuttlefish displayed a pervasive side-turning preference. Subsequent work by Alves *et al.* (2009) in a large sample ($N = 107$) confirmed the existence of a population-level bias; moreover, to find out whether or not visual perception plays a role in determining the direction of turning, cuttlefish were tested either inside the empty apparatus or with attractive visual stimuli (sand and shadow) on both sides of the T-maze apparatus. Alves *et al.* (2009), using either an empty apparatus or one in which sand and shadow were provided, found a progressive post-embryonic development of a bias to escape leftwards from 3 to 45 post-hatching days.

In vertebrates, age-dependent biases depend on functional, neurochemical or morphological asymmetries between the left and right sides of the brain (Andrew, 1991a). In cuttlefish, changes in side-turning preference during development could be linked to an asymmetrical postembryonic maturation of the brain. The optic lobes are paired structures of the central nervous system in cuttlefish (Nixon and Young, 2003). Alves *et al.* (2009) reported a correlation between side-turning bias and the size of the left and right optic lobes of cuttlefish. Cuttlefish with a right turning preference possess a bigger left optic lobe, and cuttlefish with a left-turning preference possess a bigger right optic lobe (Jozet-Alves *et al.*, 2012).

In the fruitfly *Drosophila melanogaster* a previously unknown structure has been described, near the fan-shaped body, which connects the right and the left hemisphere, i.e. an asymmetrical round body (AB) with a diameter of about 10 μm. In a sample of 2550 wild-type flies, 92.4% of individuals were found to show the AB in the right hemisphere, and the natural exceptions to this asymmetry constituted only 7.6% in the population (Pascual *et al.*, 2004). Wild-type flies presenting symmetric structures were trained to associate an odour with an electric shock: a single training cycle for short-term memory testing and five individual training sessions (15-min rest intervals) for long-term memory testing. Pascual *et al.* (2004) observed no evidence of 4-day long-term memory in wild-type flies with a symmetrical structure, although their short-term memory was intact. Asymmetrical flies had both short- and long-term memory. Thus, brain asymmetry is not required to establish short-term memory but it is important in the formation or retrieval of long-term memory in *Drosophila*.

Models based on small systems of neurons provided by invertebrates, such as the nematode *Caenorhabditis elegans*, offer a unique opportunity to address the question of how symmetrical neuronal assemblies deviate to create functional lateralization. Hobert *et al.* (2002) provided a detailed cellular and molecular perspective on left–right (L–R) asymmetry in the nervous system of *C. elegans*. In this species, 2/3 of the neurons (198 out of a total of 302) are present as bilaterally symmetrical pairs of neurons. These neuron pairs (or neuroblasts) that are initially bilaterally symmetrical choose at some stage to execute a L–R asymmetrical programme of further differentiation, in terms of migratory pattern, axonal paths or gene expression patterns.

The taste receptor neurons in *C. elegans* are an example of directional asymmetry. The ASEL/ASER neurons are the main taste receptors of *C. elegans*. ASEL and ASER are bilaterally symmetrical with regard to cell position, axon morphology, outgrowth and placement, dendritic morphology and qualitative aspects of synaptic connectivity patterns. However, three putative sensory receptors of the guanylyl cyclase class, *gcy-5*, *gcy-6* and *gcy-7*, are asymmetrically expressed in ASEL (*gcy-6, gcy-7*) and ASER (*gcy-5*), two to the left and one to the right. This asymmetry of gene expression correlates with a significant functional asymmetry of the two neurons: laser-ablation studies revealed that each of the individual neurons is responsible for sensing a distinct class of water-soluble chemicals.

It seems that invertebrates not only share the attribute of lateralization with many vertebrates but may also show some similarities in their appearance. This raises the question of whether lateralization is determined by homologous genes in invertebrates and vertebrates (Vallortigara and Rogers, 2005) or whether there has been analogous evolution of lateralized function in the two taxa. New insights into the evolution of body plans and left–right specification in Bilateria have been provided recently (Grande and Patel, 2009) and they could help in the future to solve the issue. The signalling molecule Nodal, a member of the transforming growth factor-β superfamily, is involved in the molecular pathway that leads to left–right asymmetry in vertebrates (Boorman and Shimeld, 2002) and in other deuterostomes, but no nodal orthologue had been reported previously in the two main clades of Bilateria: Ecdysozoa (including flies and nematodes) and Lophotrochozoa (including snails and annelids). Grande and Patel (2009) reported the first evidence for the presence of a nodal orthologue in a non-deuterostome group, indicating that the involvement of the Nodal pathway in left–right asymmetry might have been an ancestral feature of the Bilateria. Furthermore, this study suggests that nodal was present in the common ancestor of bilaterians and it too may have been expressed asymmetrically.

Very recently the cellular and molecular mechanisms that lead to neuronal asymmetries in the nematode *C. elegans* have been investigated and compared to the mechanisms involved in asymmetrical neural development in the zebra-fish, *Danio rerio* (Taylor *et al.*, 2010). The specification of the left and right Amphid Wing 'C' (AWC) neurons of the nematode worm olfactory system and the asymmetry in the fish epithalamus has been analysed. It has been found that both these species use interactive cell–cell communication, i.e. reciprocal inter-actions between neural cells rather than a simple linear pathway, to establish left–right neuronal identity, and this reinforces the left–right asymmetry but with different outcomes and molecular details in each species. The functional differences in morphologically identical neurons in the olfactory system of *C. elegans* are the result of gap-junctional communication and calcium influxes; whereas the neuroanatomical left–right differences in the epithalamus of *Danio rerio* are the result of morphogenic changes regulated by secreted signalling molecules.

Although it is remarkable that the two species considered share the inter-action of neurons across the midline during formation of the asymmetrical nervous system and share the inherently stochastic nature of some develop-mental pathway, results need to be interpreted with caution since the evolu-tionary gap between the 302 neurons of the worm and the estimated 78,000 neurons of the larval fish (Hill *et al.*, 2003) is considerable. The striking differ-ences in the genetic and cellular pathways underline the improbability that nematode and zebrafish lateralization arose from the same ancestral event; it is more reasonable to assume at this stage that the left–right differences in the two species have evolved by convergence (see for further discussion Chapter 3).

## 1.6 Conclusion

The widespread interest that ethologists and behavioural ecologists are now giving to left–right asymmetries in animal behaviour, and the recent links of studies of cerebral asymmetries with developmental and (as we shall see in Chapter 2) with evolutionary biology, have revitalized a field that in human neuropsychology was showing signs of diminishing interest. Hopefully, within a proper biological framework the long-lasting puzzle of the evolutionary origins and mechanisms of the left and right brain specializations can find a solution. These will be the topics of the next chapters.

# Function

## Summary

Lateralization is manifested in two main ways: (1) in individuals but with no common direction (bias) in the group or population, or (2) in individuals and in the same direction in most individuals so that the group or population is biased. The first is discussed in terms of evidence of efficiency of neural processing in a lateralized brain. The second is discussed mainly in terms of the hypothesis that population biases occur as evolutionarily stable strategies when lateralized individuals coordinate with each other. This hypothesis is supported both by recent evidence and by mathematical models.

## 2.1 Introduction

It is believed that bilateral symmetry evolved when organisms adopted an axial orientation to their direction of movement and it is usually agreed that the pathway to a bilateral nervous system led from radial symmetry. In addition to being bilaterally divided, however, the nervous system of vertebrates shows a pervasively contralateral organization in that afferent and efferent pathways cross the midline of the body so that each side of the brain connects to the opposite side of the body. Also, as we have already seen (Chapter 1), the nervous system has a certain degree of asymmetry between the left and the right sides, and this is seen in both function and structure. Before considering the function of such an asymmetrical organization let us discuss the problem of why the nervous system is organized contralaterally (Figure 2.1), given that we shall refer to such an organization almost continuously while describing experiments and observations on asymmetries in animal behaviour.

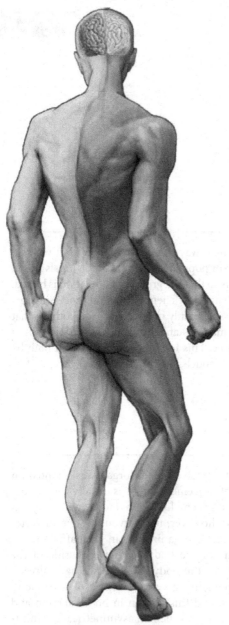

Figure 2.1 Crossing of nerves. In people and other vertebrates, nerves to and from one side of the body are linked to the hemisphere of the brain on the opposite side. As a result, each hemisphere generally controls the opposite side of the body and, apart from olfaction, sensory inputs from one side go to the opposite side of the brain. From MacNeilage *et al.* (2009).

## 2.2 Midline crossing of sensory inputs and motor outputs

An idea suggested to answer the problem pointed out the importance of unilateral stimuli eliciting a contralateral muscle response in the form of a defensive reflex (Young, 1962), the hypothesis being that preservation from damage or harm might be more effective if neural control were laterally displaced from sensory receptors and motor execution. Another traditional explanation (attributed to Ramon Y Cajal: see Braitenberg, 1984) for the crossing of visual sensory projections is that the decussation (crossing of the midline) of the optic tract is needed in order to nullify the inversion of the visual image caused by the lens in the eye, and that the pyramidal (motor) decussation follows as a consequence. However, no good evidence supports these ideas, and furthermore not all sensory inputs cross the midline to be processed in the contralateral side of the brain. The putative most ancient of the senses, olfaction, shows no decussation even though odour inputs sent ipsilaterally to the brain do affect muscles on the contralateral side of the body (i.e. the motoric side is crossed). (There could be an anatomical reason for such a lack of crossover in the case of olfaction, owing to the fact that olfactory perceptual input goes directly to the anterior tip of each hemisphere.)

An alternative hypothesis for the midline crossing of sensory inputs and motor outputs, proposed by Valentino Braitenberg in his book *Vehicles* (1984), is schematized in Figure 2.2. Imagine a very simple primitive organism with only two sensory units and two motor units. Suppose that the effects of a sensed distal

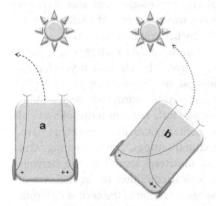

Figure 2.2 Wiring of sensory motor units. In the example, + indicates that the wheels get proportionally more torque in the vicinity of a source, and ++ indicates higher activation (more torque) of the wheel connected to the receptor closer to the source of light. Depending on the pattern of crossing, the vehicle on the left (vehicle a) orients away from the source and the vehicle on the right (b) towards it. The reverse would occur if the source is harmful (-), in this case reducing the torque of the wheels depending on vicinity.

source of stimulation (say light) were to increase, proportionally, the activity in the two motor units, as shown in Figure 2.2. Depending on the wiring between the sensory and motor units the organism would move away from (when there is no crossing of the fibres) or move towards (when there is crossing of the fibres) the distal source of excitation. The problem obviously is that we do not have any clear idea of the type of propulsion system used by primitive organisms: was it something like that of a jellyfish or something like that of a fish? Thus we still do not really know why and how the crossing over of pathways occurs. In Chapter 3 we consider the issue again, discussing the possibility that crossing may have been a consequence of initial body asymmetry.

## 2.3 Lateralization of everyday behaviour

Now, let us come back to the issue of the biological function of asymmetry of the brain. Traditionally, the study of brain and behavioural asymmetries has been the sole province of neurology and neuropsychology. Biologists have shown little interest in this topic until relatively recently. The reason for this delayed interest could not have been simply a lack of evidence for lateralization in non-human species, because topics such as the evolution of verbal language attracted considerable interest among biologists in spite of being (allegedly) a uniquely human phenomenon. Note, also, that, as mentioned in Chapter 1, behavioural evidence for lateralization in non-human animals dates back to more than 40 years ago, and neuroanatomical evidence for asymmetries in the nervous system dates back to the nineteenth century. Lack of interest among biologists was, therefore, not related to lack of evidence. What was missing was evidence that lateralization is important in a natural context, in the everyday behaviour of animals.

The fact that brain and behavioural asymmetries do matter in a natural environment is not immediately apparent from the classical psychological and neurological literature. To an ethologist or an evolutionary biologist, brain and behavioural asymmetries demonstrated in neuropsychological studies appear to be bizarre phenomena that make little sense in terms of survival and biological fitness in a natural environment. In humans asymmetries have, typically, been observed in two highly unnatural settings. The first setting takes advantage of pathology (the outcome of unilateral lesions to the nervous system) or of the side effects of a pathological condition (as in the case of split-brain patients who had undergone surgical section of the corpus callosum to reduce otherwise intractable epilepsy; Gazzaniga, 1967; Beaumont, 1982). The second setting involves sophisticated measurements of small differences observed under very unnatural conditions of stimulation (for example ultra-short presentation of visual stimuli to the left and right periphery of the visual field; Bradshaw, 1991). Understandably, this is not the sort of evidence that would impress an ethologist. And of course the same limitations apply to early work carried out with non-human animals, particularly canaries, chicks and rats, involving brain lesions or pharmacological techniques (mentioned in Chapter 1). The only exception to this state of affairs was represented by research on human handedness, which clearly has consequences and effects on everyday behaviour of human beings. Intriguingly, however, recently ethologically oriented research on other forms of asymmetry in human behaviour has begun to appear, examples including direction of kissing in airports (Güntürkün, 2003), head turning for hearing in noisy discos (Marzoli and Tommasi, 2009) and preferences of seating position in theatres and other public places (Weyers et al., 2006; Okubo, 2010).

As described in the previous chapter, we now know that a variety of asymmetries are apparent in the everyday behaviour of animals of virtually all taxonomic groups (Chapter 1). Fish keep track of companions using by preference their left eye and of a predator using their right eye; lizards, toads, baboons and wild horses are more likely to attack a conspecific competitor seen by the left eye; chickens and Australian magpies look at aerial predators using their left eye; black-winged stilts (*Himantopus himantopus*) use their right monocular visual field before predatory pecking, whereas shaking behaviour, a component of courtship displays, and copulatory attempts by males are more likely to occur when females are seen with the left monocular visual field.

The very existence of these biases is worth stressing, for it is neither obvious nor mandatory that asymmetries present in the nervous system (e.g. in neuroanatomy or neurochemistry) should be manifested as asymmetries in behaviour. Of course differences in behaviour between, say, individuals that show or do not show (or that show to a different degree) asymmetries in the nervous system may be expected. But the fact that left or right biases are apparent in behaviour raises intriguing issues about the advantages and disadvantages of possessing such overt asymmetries.

Consider a circumstance in which execution of a particular behaviour may affect in a direct and crucial way the survival of an organism, as in the case of response to predators. Animals such as fish escape predators or other noxious stimuli by turning away from them. And, in principle, all other conditions being equal, predators or other dangerous stimuli have the same probability of appearing to the left or to the right side of the potential prey. Therefore, one would expect animals to be prepared to detect and escape dangerous stimuli appearing to the right or to the left with identical ability. Surprisingly, this does not seem to be the case. Several species of fish, for example, show striking asymmetries – at the level of both the peripheral (muscles) and central (nervous system) structures – making escape more likely to occur on one particular side. This is puzzling. Why would natural selection have led to fish with such a weird and apparently ineffective strategy of responding?

We know quite well the behavioural mechanism associated with escape response in fish. This is shown schematically in Figure 2.3 (top). When a

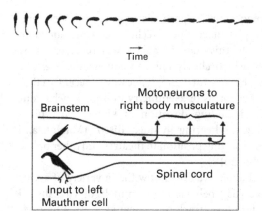

Figure 2.3 The C-start reaction of a fish (top). The arrow indicates the direction of time of the sequence. Bottom, the neural circuitry underlying the C-start reaction. Neural fibres from the Mauthner cells decussate and innervate motor neurons on the contralateral side. In several species of fish the Mauthner cells are asymmetric and produce a bias in the C-start reaction (see text).

dangerous stimulus is detected, fish react with a C-bending of the trunk, followed by a tail flip by which they gain considerable speed and acceleration. The neural circuitry underlying this 'C-start reaction' is shown schematically in Figure 2.3 (bottom). In the brainstem there is a pair of very large neurons, so-called Mauthner cells, that, after decussation, innervate motoneurons that stimulate muscles on the contralateral side of the body, thus producing the C-bending of the trunk. Fascinatingly, a variety of species of fish have been shown to be lateralized in the C-start reaction, systematically preferring bending to the left or to the right (there are differences between species but usually the left side is favoured; Heuts, 1999; Lippolis *et al.*, 2009). In addition to behaviour, the asymmetry is manifested in the left/right sizes of the Mauthner cells and their axons, as well as in the volume of muscles on the left and right side (Heuts, 1999).

## 2.4 Lateralization and speed of escape

It seems clear that there would be striking disadvantages in possessing a perceptual or a motor system that is asymmetrical to any substantial degree. Any deficit on one side would leave an animal vulnerable to attack from that side or unable to attack prey or competitors appearing on that side. It could be argued that the asymmetry may be small in magnitude. However, when the issue at hand is to escape a predator, even a small difference may have a significant effect on survival.

A possible explanation for this paradoxical design has been suggested: there could be a trade-off between speed and direction of response in the C-start reaction (Vallortigara, 2000). Let us suppose the fish is playing a sort of zero-sum game. A certain degree of C-bending of the trunk is available, say 100. A fish could decide to split its responses into two identical parts: 50 for turning to the right and 50 for turning to the left. The animal would have no bias in the direction of escape response but it might not be very fast in responding. An alternative for the fish is to be totally asymmetric, turning always, say, 100 to the right. The fish would be strongly biased and in 50% of occasions it would bump into the predator but, nonetheless, the strategy might work because the animal would be very fast in responding.

Recently, some empirical evidence suggesting that this speculative idea may be correct has been reported. Measurements by high-speed video of escape responses elicited by mechanical stimulation in the shiner perch *Cymatogaster aggregata* have revealed that strongly lateralized fish show higher escape velocities than weakly lateralized fish (Dadda *et al.*, 2010). The strongly lateralized fish show shorter latencies to respond and longer distances travelled than weakly lateralized fish.

This is just an example of one possible solution to a very general problem. The physical world is, at least at the level of macroscopic objects, indifferent to left and right. Thus, any asymmetry affecting the sensory or motor response of an organism with a basic bilateral plan of organization could turn out to be disadvantageous. Consider again the evidence we discussed that animals with laterally placed eyes might use their left and right visual fields differently, and thus that they may also perceive quite different portions of the visual world, through their left and right monocular visual fields. The problem, of course, is that any asymmetry of responding to visual stimuli could cause serious problems in some circumstances. Predators, for instance, should be detected and avoided with identical probability regardless of whether detection occurred in one or the other monocular visual field. It can be argued that species with laterally placed eyes could turn their heads to ensure that the stimulus is seen with both eyes, either by using alternatively both monocular fields to view the stimulus or by detecting it in one monocular field and then turning to use the binocular field. However, such head movements may present problems, because they can be detected by predators (or by prey) and also increase time of processing. It is likely, therefore, that some processing occurs without any lateralization in both monocular visual fields, whereas other analyses are carried out quite differently by inputs to the brain from the left and right eyes.

Summing up, it seems clear that we can expect that some advantage exists for individuals that possess asymmetrical brains and that this advantage in brain efficiency may counteract the disadvantages associated with possession of response biases. Velocity in escape responses is one example, in which non-lateralized individuals would show higher sensory threshold for the activation of Mauthner cells than would lateralized individuals. Other possible advantages can be considered.

## 2.5 Increased neural capacity

Lateralization may offer a crucial advantage by increasing neural capacity, since specializing one hemisphere for some particular functions leaves the other hemisphere free to perform other (additional) functions. This would avoid useless duplication of functions in the two hemispheres, thereby sparing neural tissue. Such an advantage is, however, somewhat counteracted by the advantages that redundancy of neural mechanism may offer following unilateral damage. For example, in humans, at least in adulthood, lesions to regions of the left hemisphere that control language production may give rise to serious impairment of speech that cannot be recovered through functions carried out by homologous regions of the other (right) hemisphere.

Levy (1969) has argued that lateralization evolved as an adaptation permitting control of the vocal apparatus unique to humans, uncomplicated by any competition between the two hemispheres. This hypothesis generated the prediction (for which there is some empirical evidence; see Levy, 1969) that people with bilateral representation of language or partial language competence in both hemispheres would perform poorly in tests of visuo-perceptual functions typical of the right hemisphere. However, the existence of asymmetry in non-human species makes it very unlikely that unilateral control of speech could have been the most important event leading to the evolution of brain asymmetry. Nonetheless, the idea that dominance by one hemisphere (or, more generally, by one side of the brain) is a way of preventing the simultaneous initiation of incompatible responses is likely to have been important in evolution, particularly for organisms with laterally placed eyes since each eye sends entirely different information to the brain hemispheres (Andrew et al., 1982; Cantalupo et al., 1995). When animals that receive different and separate views of the world have to deal with the control of medial and unpaired motor organs, such as the tongue or the beak, one way of solving the potential conflict of input in order to pursue a unitary course of action is to use inhibitory circuits in one or other side of the brain, making one side 'dominant' over the other. In birds, the fact that one side of the brain, in several circumstances, takes charge of control of behaviour is apparent from comparison of performance when they are tested binocularly and monocularly. A striking example is shown in Figure 2.4.

The figure depicts the results of some experiments in which domestic chicks were trained with both eyes in use (binocularly) to scratch on a floor covered with sawdust to find a food container (in the figure the intensity of the grey areas represents the most intensively scratched areas; see Vallortigara, 2000; Tommasi and Vallortigara, 2001). During training, the position of the food

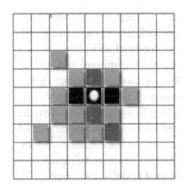 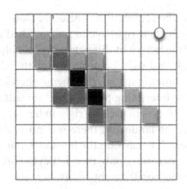

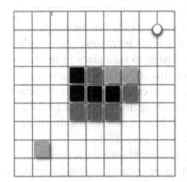 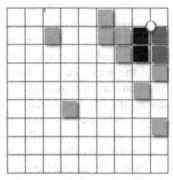

Figure 2.4 Spatial behaviour. Chicks were trained binocularly to find food by ground scratching on the floor of an enclosure. Top left: the position of the food in the centre of the enclosure was also indicated by a landmark cylinder (white circle). On testing, after displacement of the landmark into a corner, both binocular (shown at top right) and left-eyed (right hemisphere in use – shown at bottom left) chicks searched in the centre, whereas right-eyed chicks (left hemisphere in use – shown at bottom right) searched near the landmark. Note that binocular and left-eyed chicks showed similar responses, suggesting that in the normal (binocular) viewing condition the right hemisphere is in full control of this behaviour (see text and Vallortigara, 2000, for original data). The shading of the squares indicates the amount of scratching in each, and dark means more scratching.

was indicated both by a local landmark (the cylinder shown in the figure) and the central positioning within the arena. After training binocularly, chicks were tested either binocularly or monocularly following displacement of the land-mark to a corner. Chicks tested monocularly with only their left eye in use (the right eye being temporally occluded by a patch) persisted in searching in the centre of the enclosure (lower left of figure), whereas chicks tested with only their right eye in use searched near the landmark (lower right in figure). This left–right difference is probably due to the differential role played by the left and right hemispheres in attending to local (left hemisphere) and global, geometric spatial (right hemisphere) cues, described previously (Chapter 1). But the interesting point in this context is the behaviour of the chicks tested with both eyes in use (top right in figure). As can be seen, following the displacement of the landmark, there was no evidence of mixed or bimodal distribution of search behaviour: the binocular chicks mostly searched in the centre, as did the chicks using only their left eye. This shows clearly that in the usual, binocular condition the right hemisphere takes charge of control of the sensory cues to be used for spatial orientation.

## 2.6 Parallel processing

Another hypothesis that has been put forward as a possible function of lateralization is that brain asymmetry represents a way of increasing the brain's capacity by enabling separate and parallel processing to take place in the two sides of the brain. There is direct evidence for this hypothesis in both non-human and human animal species. Rogers *et al.* tested chicks on a dual task, one involving the left hemisphere in control of searching for food and pecking responses and the other involving the right hemisphere in monitoring overhead to detect a model predator (Rogers *et al.*, 2004; see Figure 2.5).

Chicks exposed to light before hatching were compared with those incubated in the dark, since the light exposure triggers the development of visual lateralization on these tasks (see Chapter 4). Lateralized (light-exposed) chicks detected the model predator sooner with the left eye (i.e. when the right hemisphere was attending to the stimulus) than with the right eye, and dark-incubated chicks showed no such lateralization (Rogers, 2000). Rogers *et al.* (2004) scored not only the response to the model predator but also the chick's ability to learn to peck at grains versus pebbles. Light-exposed, lateralized chicks avoided pecking at pebbles far better than dark-incubated, non-lateralized chicks and they were more likely to detect the model predator. Once they had detected the model predator, the dark-incubated chicks were more disturbed by it and more distracted from searching for food (Wichman *et al.*, 2008). As a control, the dark-incubated chicks were tested on the pebble–grain test without presenting the model predator: in this case they had no difficulty in learning to find the grain and avoid pecking at the pebbles. Hence, the dark-incubated chicks had their greatest difficulties when they attended to the two separate tasks simultaneously. These results were subsequently confirmed in a species of fish tested in a similar dual task (e.g. Dadda and Bisazza, 2006). Moreover, research with humans suggests that the dual nature of the

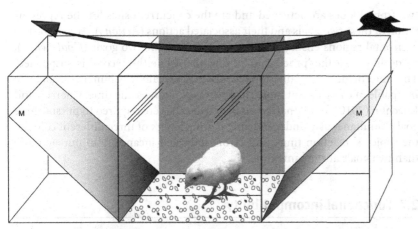

Figure 2.5 Dual task. Schematic representation of the dual task in which chicks discriminate between grains of food and pebbles on the floor while simultaneously attending to a simulated predator overhead. The chick is videotaped from above, and the mirrors (M) aid accuracy of scoring since the chick can be seen from three angles. From Rogers *et al.* (2004).

brain associated with lateralization of function may explain some of our limitations in the capacity to make decisions.

The part of the human brain that controls motivation to pursue a goal can divide its attention between two tasks. The left half devotes itself to one task and the right half to the other. This division of labour allows us to multi-task. The frontopolar cortex, an area at the very front of our brains, drives our ability to do more than one thing at a time. It allows us to pursue two different goals simultaneously, holding one at the ready while we work on the other. Just behind the frontopolar cortex lies the medial frontal cortex (MFC), an area which is involved in motivation. It controls our pursuit of multiple goals, according to the rewards we expect from them. When humans pursue a single goal, such as pairing letters appearing successively on a screen in order to obtain a reward, the task is coded, according to the expected reward, symmetrically in regions of the left and right medial frontal lobes simultaneously (Charron and Koechlin, 2010). When, however, the subjects are facing two tasks at the same time, for example pairing capital letters in one task and small letters in the other, the left region of the brain codes the rewards driving one task while the right region codes those associated with the other task. It seems as if each frontal lobe is pursuing its own goal. But how does the human brain coordinate the pursuit of goals belonging to the left and right frontal lobes? Charron and Koechlin (2010) have shown that the most anterior prefrontal regions deal with coordinating the execution of these objectives. When a human subject pursues two independent goals A and B (*Goal A/Goal B*), the two frontal lobes are activated and are the concurrent sites for the representation of these two goals and their associated actions (*Action A/Action B*). The prefrontal regions ensure coordination between the two goals (*Goal switch*), by dealing with the 'processing' of one goal while the second is suspended. This 'interhemispheric' division of work may explain why humans seem to be incapable of carrying out more than two tasks at the same time. Charron and Koechlin (2010) found that, when their subjects were required to pursue three goals simultaneously, independently pairing letters of three different colours, the subjects' reaction times and errors indicated incapacity to pursue more than two goals at the same time.

## 2.7 Functional incompatibility

Hints about further biological functions of lateralization come from the way in which different cognitive abilities seem to be assigned to the left and right sides of the brain (see Chapter 1). It has been suggested that, from a computational point of view, segregation of functions in the separate halves of the brain may represent a solution to a problem of 'functional incompatibility' (Vallortigara *et al.*, 1999). When assessing a novel stimulus, an ability needed quite commonly by even the most primitive vertebrates, two different types of analyses must be carried out. First, the organism must rely on previous comparable experiences to estimate the degree of novelty of the stimulus, and to do so it must recall stored memories and then elaborate on them for future use. This process requires attention to the unique features of an event or a stimulus (i.e. the variance, outside the range expected from previous experiences). Second, certain appropriate cues, based on past experience and/or on phylogenetically based instructions, must be used to try to assign the stimulus to a category, and so to decide what sort of response (if any) should be given. Categorization must be made on the basis of selected stimulus properties, despite variation in many other properties.

All this is reminiscent of the functional incompatibility among logical demands, a condition hypothesized to underlie the evolution of multiple systems in biology (see e.g. Sherry and Schachter, 1987 for the evolution of multiple systems in memory). Functional incompatibility means that the mechanisms adequate for the solution of one problem are incompatible with those needed to make the solution of another problem.

An example is the trade-off between sensitivity to light and spatial resolution in the visual receptors in the retina. Why should there be a dual system of cones and rods in the retina? Probably because of functional incompatibility, since fine resolution by cones is achieved at the cost of reduced ability to detect changes in average light intensity. In order to be able to detect small changes in average light intensity, signals from single receptors should converge and add together on single neurons; in contrast, in order to be able to show fine spatial resolution, very limited convergence of signals from receptors to retinal neurons is needed. A similar argument applies to cognitive mechanisms of categorization and response to novelty. In order to categorize events (or stimuli), the organism must recognize, and memorize, those features of an experience (or of a stimulus) that recur in different episodes (or stimuli) and, at the same time, ignore or discard unique and idiosyncratic features that do not recur and thus are not essential to remember. This form of processing responds to invariance. The selective attention that results is one of the brain's main functions, as it enables the smooth, and eventually automatic, execution of skilled motor behaviour, performed in response to certain invariant features of episodes. In contrast, to detect novelty and to build up a detailed record of episodic experiences to which the organism must attend, the contextual details must be used to mark individual experiences uniquely, i.e. to recognize variance across episodes rather than invariance. These two types of processing need to be carried out simultaneously if an animal is to be able both to detect novelty (which can depend on detecting features that cannot be predicted in advance) and to categorize objects and events so that they can be responded to appropriately by 'practised' responses and skills. These two functions may be incompatible, and this fact may provide the clue to hemispheric specialization (Vallortigara et al., 1999). Processing of information about invariance and variance might best be handled separately by the hemispheres, variance by the right hemisphere and invariance by the left hemisphere. An example is provided by studies of individual recognition in chicks (reviewed in Vallortigara and Andrew, 1994a). The young chick is facing the problem of responding to social partners in general and at the same time to recognize, within the category of social partners, particular individuals. The evidence from experiments involving free-choice tests in which chicks can associate with companions (cage-mates) or strangers suggest that although the left hemisphere is certainly capable of strong response to social partners, it is not competent to distinguish between individuals. The right hemisphere in contrast excels in individual recognition abilities. A plausible explanation for that is that the left hemisphere tends to build up representations of categories of 'social partner', using chiefly those invariant features that belong to all members of each category. The right hemisphere, on the other hand, is particularly suited to build up representations based on past records of specific experiences that are as complete as possible. The assessment of novelty is clearly central to the identification of a stranger. As it is likely to be difficult or impossible to choose in advance the features that may be different in any particular stranger, it is necessary to have available full descriptions of familiar fellows, if strangers have to be distinguished from companions. These in turn must be compared as fully as possible during analysis with the new object, if novelty is to be assessed.

## 2.8 Enhanced cognition

So far we have considered evidence that asymmetry may enhance efficiency of a brain. One may wonder whether laterality enhances cognition in general. There is some evidence that this may be so, but the issue needs some scrutiny. For instance, it has been shown that strongly lateralized parrots are more adept at discriminating between food versus non-food items and in solving a complex problem than are non-lateralized parrots (Margat and Brown, 2009). Interestingly, it was the strength of lateralization but not the direction (left or right) that was the primary predictor of parrots' performance. The same has been shown very clearly in foraging for termites by wild chimpanzees (see Figure 2.6). Both hands are used by chimpanzees in termite fishing, one to hold the twig used as a probe and the other to act as a stabilizer across which the twig covered in termites is rubbed when the chimpanzee eats them. Although (see Chapter 1) there is evidence for population-level handedness in some tasks in chimpanzees, and some evidence for a population-level hand preference in termite fishing among wild chimpanzees (Lonsdorf and Hopkins, 2005), here it is individual lateralization that counts. Some individuals preferentially use the same hand to probe and the other to stabilize the twig (lateralized), whereas others vary which hand is used for either purpose (ambidextrous). McGrew and Marchant (1999) measured the efficiency of termite fishing by the chimpanzees and found that individually lateralized chimpanzees, irrespective of the direction of their lateralization, gathered more prey for a given amount of effort than did ambidextrous chimpanzees. Thus, individual lateralization clearly suffices to confer an advantage without any need for alignment of lateralization at the population level (a point to which we shall return soon).

In pigeons, Güntürkün et al. (2000) found that the stronger the left versus right eye lateralization in discriminating grains from pebbles, the more successful were the birds in selecting grains when they were tested binocularly. Again, these results show that stronger lateralization increases efficiency of performance. However, it is important to stress that arguing for computational advantages associated with having a lateralized brain is not the same as arguing that the more lateralized a brain, the more efficient it will be. This is so for two reasons. First, the relationship between brain asymmetry and cognitive performance is task-dependent (Boles et al., 2008). Second, even in a particular task, the relationship between lateralization and cognitive performance is not always linearly positive, as it is in the case of pigeons discriminating grains and pebbles. Hirnstein et al. (2010) provided some empirical data on this issue. They found that in human subjects the relationship between lateralization and

Figure 2.6 'Termite fishing' by chimpanzees. Source: http://www.wildchimpanzees.org/ press/photo.php. McGrew and Marchant (1999) found that individuals with a stronger hand preference forage more efficiently but this does not depend on which hand is preferred. (For behaviours showing population-level preference in hand use see Chapter 1.)

cognitive performance in two behavioural tasks (a left lateralized word matching task and a right lateralized face-decision task) was best described by an inverted U-shaped curve. The optimal degree of asymmetry was task-specific but, in general, a moderate degree of asymmetry was associated with the best cognitive performances in both tasks. Cognitive performances, therefore, deteriorated towards both extreme ends of asymmetry, i.e. in symmetrical and in very strongly asymmetrical individuals.

The conclusion we have reached so far is that it seems that there are several advantages associated with possession of a moderately asymmetrical brain. Also, given that these advantages are not mutually exclusive, it is likely that they could all have had a role in evolution. In several respects, a (slightly) asymmetrical brain would function better than a symmetrical one. These 'computational' advantages can very probably counteract the disadvantages associated with possession of biases in overt behaviour.

## 2.9 Population-level biases

There is a problem, however, in viewing lateralization solely as an advantage in the computational abilities of the brain. Most of the asymmetries we have described so far are 'population-level' or 'species-level' asymmetries, i.e. asymmetries showing a similar direction in more than 50% of the population. In biological terms these are 'directional' asymmetries, quite distinct from asymmetries occurring at the 'individual level', i.e. asymmetries in which 50% of the individuals favour the left and 50% favour the right (also referred to as 'antisymmetry'; see Denenberg, 1981; Palmer, 1996).

Although the hypothesis that lateralization increases the brain's efficiency may explain the presence of individual lateralization, it does not, in itself, explain the alignment of the direction of lateralization at the population level, because individual brain efficiency is unrelated to how other individuals are lateralized. Why, therefore, do animals possess brain lateralization at the population level? Would it not be simpler for brain lateralization to be present in individuals without any specification of its direction (i.e. with a 50:50 distribution of the left and right forms in the population)?

One possible explanation is that population-level asymmetries are linked to some underlying asymmetry of living matter, such as the helical coiling of amino acids. It is also possible that alignment of the direction of lateralization at the population level may be a mere by-product of genetic mechanisms that specify left–right differences in structure of the body and brain (McManus, 2002): that is, genes could specify not only strength of lateralization but also its direction. The point here is whether genetic encoding of the presence of an asymmetry (with a certain strength) is mandatorily accompanied by genetic specification of its direction. Evidence suggests that it is not, because we know that sometimes asymmetries – both in the nervous system and in behaviour – can be observed at the individual and not population level (reviewed by Bisazza et al., 1998). Moreover, current genetic evidence suggests that genes rarely determine the direction of asymmetry, but rather tend to influence whether or not an asymmetry will be expressed. Even more important, we know that artificial selection for the strength of lateralization without inheritance of direction is perfectly possible. This has been shown to be the case, for instance, in mice by Collins (1985). The 'degree' of laterality for pawedness in mice, defined as the absolute difference between preference to reach for food using the left or right paw, could be changed by selective breeding, whereas the direction of paw preference could not be changed (Collins, 1985, 1991). Collins et al. (1993) demonstrated that the differences between the genetic lines that had been bred for strong versus weak laterality were not due to differences in genetic heterogeneity. A low degree of laterality, i.e. an equal preference for using the left or right paw, was associated with reduced asymmetry of the brain hemispheres (Collins, 1985). Mice selected by Collins for a high degree of laterality showed more brain asymmetries than mice selected for a low degree of laterality (Ward and Collins, 1985). Strength and direction of asymmetries are not necessarily genetically linked. Genes could encode for the mere presence of brain asymmetry leaving at random the specification of its direction (thus, individuals will be asymmetrical but the population will comprise one-half left asymmetric and one-half right asymmetric).

It is noteworthy that the case described for pawedness in mice is not the rule. It provides evidence that genetic specification of the presence of asymmetry without genetic specification of its direction is possible. However, in other cases there is clear evidence of genetic inheritance of both strength and direction of asymmetries. This is the case, for instance, for the teleost fish *Girardinus falcatus*. These fish show population-level lateralization in a detour test in which they can turn rightward or leftward when facing a dummy predator visible behind a barrier. Although at the population level fish turn left approximately 65% of times and right 35% of times, there is a wide individual variation in the population. When males and females with the same score at the detour test are paired and their progeny tested with the same behavioural procedure, a strong correlation is apparent between scores of parents and of their offspring (Bisazza *et al.*, 2000, 2001). Heritability exceeds 0.5 – a notably high value in view of the nature of the character measured. The same result is obtained whether the progeny were raised with their parents or whether the progeny were separated from parents at birth, strongly suggesting the existence of genes controlling the direction of brain lateralization. Thus, the problem is to understand why in some cases genes code for an alignment of lateralization at the population level and in other cases they do not.

A further puzzle is that the alignment of the direction of the asymmetries in the majority of individuals (population-level or directional asymmetry) may even be disadvantageous, as it makes individual behaviour more predictable to other organisms. If most of the fish of a population C-started leftward when encountering a predator (as described above), the predator could learn about the bias of its prey and exploit it during prey catching. The same disadvantage would not hold if prey were lateralized only at the individual level. The logical conclusion is that there must have been important advantages that maintained directional asymmetry in spite of its potential disadvantage.

If the presence of brain asymmetry at the individual level is a sufficient condition to produce advantages in terms of brain efficiency, and if there are disadvantages in aligning the direction of asymmetries at the population level, why has population-level lateralization emerged at all? Even more puzzling to explain is the fact that a certain number of individuals in a population show reversed lateralization. In animal populations the proportion of individuals with reversed lateralization ranges from approximately 10% (as in the case of human handedness; McManus, 2002) to approximately 35–40% (as in the case of pawedness in toads; Bisazza *et al.*, 1996). Why does this proportion vary? It could be due to environmental effects on development, to genetic variation or, most likely, to both of these influences interacting.

Corballis (2006) has argued that the balance between individuals with asymmetry and those with symmetry is maintained through a heterozygotic advantage. Some genetic models of human handedness (McManus, 1999; Annett, 2002) posit one or more 'directional' (D) alleles that cause right-handedness, and one or more 'chance' (C) alleles that cause left- or right-handedness at random. A population with a majority of right-handers and a minority of left-handers can be maintained, in these models, if DC genotypes have higher fitness than CC and DD genotypes (heterozygotic advantage). The idea would be that individuals carrying the gene(s) for cerebral asymmetry in heterozygosis may have a slightly greater biological fitness than those carrying the gene in the homozygotic condition. For example, Annett (2002) has shown that left-handers may be over-susceptible to reading disorders and extreme right-handers may have deficits in spatial or sporting activities. One problem with the hypothesis of the heterozygote advantage, however, is that it is not clear *why* heterozygosity might be more adaptive than homozygosity. McManus (2002) has put forward the hypothesis of 'random cerebral variation' according to which beneficial combination of modules may occur more commonly in DC individuals, and in general an increased variability should be expected among left-handers. Yet, such variability would not necessarily produce an increase in biological fitness in individuals: random variation may well produce a non-beneficial combination of cerebral modules.

We believe that population-level lateralization may be instead better explained by a simple and straightforward idea: sometimes, what is advantageous for an (asymmetrical) individual to do depends on what the other (asymmetrical) individuals of the group do. The hypothesis suggests that there may be 'social' constraints that force individuals to align their asymmetries with those of the other individuals of the group (Vallortigara and Rogers, 2005). More technically, it could be argued that the alignment of the direction of behavioural asymmetries in a population can arise as an 'evolutionarily stable strategy' – a concept developed by the evolutionary biologist John Maynard-Smith (1982) in the context of theory of games and agonistic behaviour – when individually asymmetrical organisms must coordinate their behaviour with that of other asymmetrical organisms (Ghirlanda and Vallortigara, 2004; Vallortigara and Rogers, 2005). According to this view, the evolution of brain lateralization would have occurred in two steps: first, individuals became lateralized at the individual level because of computational advantages given to the machinery of the brain by an asymmetrical functioning; then individually lateralized organisms aligned the direction of their asymmetries when they started to interact in ways that made their asymmetry relevant to each other's behaviour.

The hypothesis has been explored mathematically in a game-theoretical model that considered group-living prey subjected to predation (Ghirlanda and Vallortigara, 2004; Vallortigara, 2006c). The model took into account predators and group-living prey meeting in contests in which the prey have two lateralization strategies available, 'left' and 'right', and it assumed that, when a predator attacks, lateralization can affect a prey's probability of escaping in two ways. On the one hand, individuals in large groups have a lesser risk of being targeted by predators (the so-called 'dilution' of predation risk; Foster and Treherne, 1981). This favours individuals who tend to escape in the same direction as the majority, thus promoting the same direction of lateralization across the whole population. On the other hand, given that predators may learn to anticipate prey escape strategies, individuals who escape in a different direction from the majority may surprise predators and survive predation attempts more often. This tends to favour populations in which left- and right-lateralized individuals are equally common. The model shows that population-level lateralization can emerge spontaneously under certain conditions. This is shown in Figure 2.7.

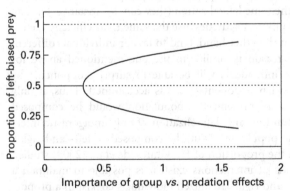

Figure 2.7 Game-theoretical model of left–right proportions in a group. Equilibrium proportion of left-type prey in a group-living species is presented as a function of the relative importance of social factors (see text for details). Solid lines: stable equilibria; dashed lines: unstable equilibria. As can be seen, when the importance of social factors is low (i.e. in solitary species) stable equilibria are apparent with 50% left-biased and 50% right-biased individuals (individual-level lateralization) up to a point at which, when sociality increases, an uneven distribution of left- and right-biased individuals should be apparent.

When the value in the abscissa, which represents the relative importance of the two factors, is small, the only stable population (solid lines) consists of left- and right-type prey in equal numbers (individual-level lateralization). This would correspond to situations in which lateralization mediated effects of group living on probability of escaping are small, as in the case of solitary prey or in the case of those kinds of lateralization that do not influence group cohesion. When the value in the abscissa becomes larger, stable populations consist of left- and right-type prey in unequal numbers (because the model does not assume any intrinsic benefit of left or right lateralization, there are always two possible solutions: one with a majority of left-type prey and one with a majority of right-type prey). The intuitive content of such a situation is that the majority of prey gain protection by keeping together but pay a cost because predators are better at handling them. A minority of prey manages to enjoy the same probability of escaping by trading off protection from the group against an advantage in the face of predators.

## 2.10 Agonistic versus synergistic interactions

The model has been extended to lateralization in purely intraspecific interactions (Ghirlanda et al., 2009). Assuming that individuals of a group engage in both antagonistic (competitive) and synergistic (cooperative) interactions, an individual's payoff depends on its success in interactions, which is a function of how common its lateralization is in the population. Synergistic activities tend to favour individuals with the same lateralization (they can, for instance, have an easier time coordinating physical activities, use the same tools efficiently, etc.). Antagonistic activities, on the other hand, tend to favour individuals different from the majority. The reason is similar to the one mentioned above for predation: minority-type individuals will be able to surprise opponents by adopting behaviour to which opponents are less accustomed. Thus, if only synergistic interactions were present, the population would be composed entirely of individuals with the same lateralization. If only antagonistic interactions were present, the population would be composed of left- and right-lateralized individuals in the proportion of 1:1. Ghirlanda et al. (2009) studied whether, when both kinds of interactions exist, it is possible to maintain a population in which left- and right-lateralized individuals coexist in a proportion different from ½, and how such a situation is influenced by model parameters. Mathematical analysis demonstrates that populations consisting of left- and right-type individuals in unequal numbers can be evolutionarily stable, based solely on strategic factors arising from intraspecific interactions.

The model makes an interesting prediction, namely that the frequency of the minority type depends on the balance between the fitness contributions of antagonistic and of synergistic interactions. When antagonistic interactions are more important for individuals' fitness, we expect the minority type to be more common. Likewise, when synergistic interactions are more important, we expect the minority type to be less common. Faurie and Raymond (2004) reported data in agreement with the model, showing, very surprisingly, that the frequency of left-handers in eight traditional societies is strongly correlated with the proportion of homicide (discussed also later in this chapter and in Chapter 3).

Although it is mathematically conceivable that stable equilibria of left- and right-type organisms in unequal numbers can arise when the fitness of each lateralized individual depends on its aligning with the direction of bias of the majority of the individuals of the group, this is merely a theoretical possibility. Mathematically the theory is sound, but is there any empirical evidence that supports it?

Some empirical evidence for the theory of population-level lateralization as an evolutionarily stable strategy is provided by the finding that left-handed humans have an advantage in sports involving dual confrontations, such as fencing, tennis and baseball, but not in non-interactive sports such as gymnastics (Raymond et al., 1996). The advantage does not arise from the well-known association between use of the left hand and direct control of it by the more visuo-spatially talented right hemisphere (McManus, 2002). Instead, the advantage is a frequency-dependent one: left-handers are relatively uncommon, so both left- and right-handers are less familiar with this category of competitor. Analyses of cricket have similarly shown that the frequency of left-handers in this sport is best explained by frequency-dependent selection mechanisms (Brooks et al., 2004).

As already mentioned, Faurie and Raymond (2004) have recently shown that the frequency of left-handers is strongly and positively correlated with the rate of homicides across traditional societies: ranging from 3% in the most pacifistic societies to 27% in the most violent and warlike. The interpretation of this finding would be that the advantage of being left-handed should be greater in a more violent context, which should result in a higher frequency of left-handers. In the absence of any selection pressure, the resulting equilibrium should be a 1:1 ratio of right- to left-handers (which has never been observed in any human population).

Two theoretical points need to be stressed. First, the theory of directional lateralization as an evolutionarily stable strategy should not be confused with theories arguing for a socio-cultural origin of lateralization (see review in McManus, 2002). Of course, there could be group pressures for conformity in asymmetric functions. In humans, for instance, manufactured objects, such as scissors, are typically made for the convenience of right-handers. However, this seems to be the consequence of the existence of directional asymmetries in hand usage rather than the cause of it. However, the theory of directional lateralization as an evolutionarily stable strategy is based on the assumption that there are genes for left–right asymmetries and that some of them code for the direction of asymmetry. Some of these 'directional' genes have been found in the zebrafish model (see e.g. Regan *et al.*, 2009; Roussigné *et al.*, 2009).

A second point to stress is that the hypothesis holds only because asymmetries in the brain are manifested in overt behaviour as left–right biases. If asymmetries in the brain are without any apparent effect in the left–right behaviour of organisms (and certainly these sorts of asymmetries do exist), according to the evolutionarily stable strategy hypothesis, no selection pressures for aligning the direction of asymmetries among different individuals would arise.

## 2.11 Testing the hypothesis about evolution of lateralization as an evolutionarily stable strategy

Even though the hypothesis of the evolutionarily stable strategy of lateralization is difficult to test in organisms living now, it does make the quite straightforward prediction that, in origin, 'social' organisms should have started to be lateralized at the population level, whereas 'solitary' organisms retained lateralization at the individual level only. This prediction obviously refers to the conditions at the origin of a very complex evolutionary trajectory. We would envisage remote ancestors with a bilaterally symmetrical organization of the body and ones in which lateralization first appeared at the individual level, because that would enhance their nervous system efficiency. As a result of the presence of lateralization in the brain (and certainly in other parts of the body), lateral biases in behaviour may have become apparent during the course of evolution. When these individually asymmetric animals started to interact to each other in a manner for which their lateral biases would matter, then selection pressures may have started to work in favour of genes that promoted the alignment of lateralization at the population level.

The problem is that when we are considering modern, current-living vertebrates, arguing that a species is entirely solitary is very difficult. Many living vertebrate species that are considered today to be solitary are derived from ancestors that were social. Hence, the extant species may have retained population-level rather than individual-level asymmetries for this reason. A similar point can be raised with respect to ontogenetic development. For instance, anuran amphibians exhibit relatively poor sociality, except in their juvenile stages: in several species, tadpoles show aggregative behaviour based on kin and familiarity, and population-level lateralization has been observed in this behaviour (Bisazza *et al.*, 2002). It is thus plausible that directional asymmetries in the relatively solitary adults are retained from the juvenile stages.

It is also apparent that some forms of asymmetries that are unlikely to have been directly selected as evolutionarily stable strategies in social contexts (say, limb usage in toads, e.g. Bisazza *et al.*, 1996) could have evolved as population-level biases as by-products of other biases that have, in fact, evolved as evolutionarily stable strategies. In other words, it is possible that, when an asymmetry aligned to produce a directional bias in the population, other asymmetries that did not require any alignment at the population level because they were irrelevant to any social interactions may have become organized as directional simply because a directional organization in the two sides of the brain already existed.

All this is not to say that the theory could not be falsified, but rather that evidence for solitary current-living species showing directional asymmetries would not necessarily be evidence against the theory (directional asymmetries in such species could represent the remnant of social ancestors). On the other hand, evidence for social current-living species showing *only* individual-level asymmetries would disprove the theory.

Some tests of the hypothesis that directional asymmetry should be found only in cooperative, social species can be carried out on certain current-living species in which the distinction between solitary and gregarious behaviour can be defined quite sharply with respect to at least some aspects of behaviour and in which it is likely that no major changes in sociality have occurred in evolutionary terms. A case in point is anti-predatory behaviour of species of fish that shoal versus those that do not. Shoaling in fish is a way of gaining protection against predators and, given the obvious advantages of shoaling, it seems unlikely that extant species that do not shoal derive from ancestors that did shoal. Thus comparison of species that do and do not shoal may provide a valuable test for the theory. Consequently, Bisazza *et al.* (2000) investigated whether shoaling in fish was associated with a population bias to turn in one direction (either left or right) when faced by a barrier of vertical bars through which a model predator could be seen. The shoaling behaviour of the species was determined either on the basis of previous ethological literature or by directly testing it: groups of fish were placed in a tank together and an index of their proximity to each other was determined. Twenty species of teleost fishes were studied. Some species were found to be gregarious (i.e. to shoal) and all of these were found to be lateralized for turning bias at the population level; some other species were found to be non-gregarious (i.e. not shoaling) and most of them were found to be lateralized at the individual level but not at the population level. Hence, the results supported the hypothesis (see also Vallortigara and Bisazza, 2002).

Comparison of lateralization in social and non-social insects may provide an even more powerful test. Although lateralization in invertebrates may not be related to lateralization in vertebrates in any evolutionary sense, the social pressures associated with the need to coordinate asymmetric behaviours would hold irrespective of whether lateralization in vertebrates and in invertebrates represents homology (common ancestry) or homoplasy (convergent evolution).

Among Hymenoptera closely related species have evolved either sophisticated eusociality or maintained solitary behaviour. Anfora *et al.* (2010) tested behavioural and electrophysiological lateralization of olfactory responses in two species of the superfamily Apoidea, the social honeybee, *Apis mellifera* L. (Fam. Apidae), and the solitary mason bee, *Osmia cornuta* (Latreille) (Fam. Megachilidae). The common name, mason bees, derives from their habit of making compartments with thick mud walls in hollow reeds or holes in wood made previously by wood-boring insects. Unlike honeybees, mason bees are solitary: every female is fertile and makes its own separate nest. Mason bees do not produce honey or wax and there are no worker bees in these species. Males leave their maternal nest before the females, and compete for mating by waiting for the female at the nest entrance. Mating occurs immediately after the female leaves her nest and then she spends two to three days feeding on flowers before starting a new nest, in which she deposits a mass of pollen and nectar and on which she lays an egg (Nepi *et al.*, 2005).

Lateralization in mason and honeybees was tested by measuring their ability to recall a learned odour using the left or right antenna. Recall of the olfactory memory one hour after training to associate an odour with a sugar reward, as revealed by the bee extending its proboscis when presented with the trained odour, was better in honeybees when they used their right antenna than when they used their left, confirming previous results obtained in the same species (Letzkus *et al.*, 2006; Rogers and Vallortigara, 2008). Hence, honeybees show population-level lateralization. No such asymmetry was observed in mason bees. Consistent with this species difference, electroantennographic responses to a floral volatile compound and to an alarm pheromone were higher in the right than in the left antenna in honeybees but not in mason bees (Figure 2.8). Although the mason bees showed no population-level lateralization, they did show evidence of individual-level lateralization in the electroantennographic responses: individual mason bees showed significantly stronger responses either with the right or the left antenna, without any alignment of lateralization in the majority of individuals. Again, these data fit nicely with the hypothesis linking directional asymmetry with social behaviour.

## 2.12 Strength of lateralization

Changes in the strength of occurrence of behavioural asymmetries that depend on ecological factors, such as degree of predatory pressure, can be predicted. In fact, Brown *et al.* (2004) found that populations of the same species of fish (*Brachyraphis episcope*) collected from parts of the stream where predation was high showed strongly lateralized behaviour, whereas fish from low predation areas showed no evidence of lateralization. It could be that shoaling was

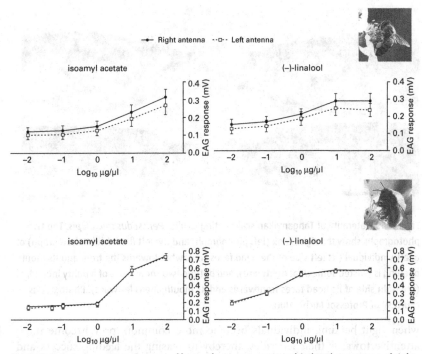

Figure 2.8 Antenna responsiveness of bees. Electroantennographic (EAG) responses of right and left antenna of *Apis mellifera* (upper graphs) and *Osmia cornuta* (lower graphs) to isoamyl acetate and (−)-linalool at five different doses. The dotted lines with empty squares show responses of the left antenna, whereas unbroken lines with black squares represent the right antenna. See Anfora *et al.* (2010).

performed more frequently by the fish experiencing higher levels of predation but this was not assessed in the study.

It is also interesting to observe that in some cases a stable polymorphism may evolve under frequency-dependent selection not in the form of directional laterality but in the form of fluctuations of left- and right-phenotypes over periods of time. A case in point is that of scale-eating fish. Hori (1993) first reported such a phenomenon in *Perissodus microlepis*, a specialized scale-eating fish from Lake Tanganyika that eats scales off the posterior flanks of larger fish. The mouths of these scale-eating fish bend to one side of the head (Figure 2.9), which allows them to strike from a more posterior orientation, so making them less visible to prey.

Mouths bend to the right in some individuals and to the left in others (bimodal frequency distribution). The frequencies of right- and left-bending individuals appear to vary cyclically around 50:50 over time. This suggests that negative frequency-dependent selection is maintaining this polymorphism:

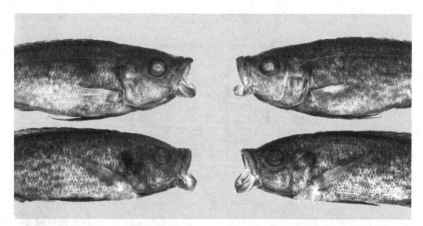

Figure 2.9 Laterality of Tanganyikan scale-eating cichlid, *Perissodus microlepis*. Top two photographs show the right flank (left photograph) and the left flank (right photograph) of a lefty individual (its left side of the head faces somewhat towards the front and its mouth opening is twisted somewhat rightward), and bottom two show those of a righty individual (the right side of its head faces frontwards and its mouth opens leftward). Photographs courtesy of Professor Michio Hori.

when right-bending individuals become more common, prey become more attentive towards their left sides, thereby increasing the feeding success and therefore the fitness of the rarer, left-bending individuals. At present, no convincing genetic models explain this asymmetry. However, there is increasing awareness that behavioural asymmetry may amplify morphological asymmetry, very probably by effects on their development (Palmer, 2010). This explanation is suggested by the fact that the frequency distribution of mouth asymmetry is not bimodal in juveniles that have not yet begun to feed on scales. It seems, therefore, that scale eating itself may affect morphological asymmetry later in life (Stewart and Albertson, 2010). Moreover, mouth asymmetry of fish forced experimentally to attack prey from their non-preferred side tended to become less pronounced over time, whereas it increased in those allowed to attack from their preferred side (Van Dooren *et al.*, 2010). Finally, the stronger the lateralized behaviour of individual fish, the more pronounced the mouth asymmetry (Van Dooren *et al.*, 2010).

## 2.13 Future ways of investigating the evolution of population-level lateralization

Another way to test the hypothesis about population-level lateralization depending on social interaction would be to check for the effects of cooperation and competition on lateralization in different species. The tactic of coordinated anti-predator behaviour provides an example of cooperation, since the behaviour of each individual is affected by similar behaviour displayed by the majority of individuals in the group. An individual's chance of survival is thus dependent on its ability to conform to the rest of the group. However, competition for food also increases with group size and therefore, in some circumstances, it may pay to behave differently from the rest of the group. Group foraging behaviour, therefore, may provide an example of competition and no or little population-level asymmetry. A correct balance between foraging and anti-predator behaviour would be essential if individuals are to maximize survival chances, and such a balance would predict variation in the pattern of lateralization observed at the population level in a wide range of species.

In conclusion, whether or not the evolutionarily stable strategy theory proves correct as a general explanation for the origins of directional lateralization, it certainly has the merit of providing a bridge for the previous disparate approaches of neuropsychology and evolutionary biology to research on cerebral lateralization. As we shall see in the final chapter, an evolutionary framework for the study of lateralization may provide novel insights even to the more classical approach focused on higher cognitive processing in humans and other animals.

# Evolution

## Summary

This chapter discusses the evolution of lateralization in vertebrates and their ancestors. Vertebrate asymmetry was dominated from the start by extraordinary bodily asymmetry, which determined the course of evolution of nervous system asymmetries. Modern Echinoderms (starfish and sea urchins) and all chordates (the group to which vertebrates belong) came from an ancestor that had extreme right–left asymmetry. Evidence from invertebrates, discussed later in this chapter, suggests that the possession of paired sense organs (including sensory inputs from paired appendages, as well as from paired eyes) has sometimes been sufficient to allow the evolution of lateralization of functions of the central nervous system. The evidence takes us far from the earlier notion that hemispheric specialization evolved in humans about 2.5 million years ago along with language, handedness and tool using. Although important steps in human evolution, discussed here, involved brain lateralization, they were shaped by pre-existing asymmetries, rather than appearing *de novo*.

## 3.1 Origins of asymmetry in chordates

Ancestral chordates lived in a marine environment (in the Cambrian/ Precambrian periods) very different from any present today. Food was available as tiny algae near the surface of the sea and as organic remains that had sunk to the sea floor; both were exploited by animals. The rarity of deep burrowing forms of life (Bambach *et al.*, 2007) meant that accumulation of edible particles in the deposits on the sea floor was greater than now. The structure of food webs ('what ate what') reveals that, because efficient predators were relatively scarce (albeit not non-existent, as shown by the recent discovery of a Cambrian, arthropod predator; Paterson *et al.*, 2011), many organisms did well at this time without the need for very effective defences against predators or the ability to flee (Bambach et al., 2007).

Fine particles were gathered by many life forms. Some fed by currents generated by ciliated tentacles, which carried food particles to the mouth ('filter feeding'); temporary or permanent attachment by a stolon (stalk) was possible during feeding (Caron et al., 2010). One group, including our ancestors (chordates) and the ancestors of echinoderms (e.g. starfish, sea urchins), carried out filter feeding, using pharyngeal gill slits (Smith, 2005). Currents set up by cilia on the gill bars entered through the mouth and left through the gill slits.

An animal living today, the lancelet, *Branchiostoma*, known formerly as *Amphioxus*, provides key evidence for early chordate evolution (Figure 3.1). The basic organization of the lancelet body is vertebrate-like: the same 14 Hox genes determine the linear organization of the body in lancelets and vertebrates (Swalla and Smith, 2008). The lancelet's central nervous system is a hollow dorsal structure with detailed resemblances to the vertebrate central nervous system. A notochord is present, and the segmental myotomes of the trunk allow fish-like swimming by side-to-side movements of the trunk. One important difference between lancelets and vertebrates is the absence of paired eyes in lancelets (discussed further below).

Lancelets have effective locomotor musculature but, since adult lancelets rotate when swimming, they are not capable of pursuing prey by swimming. The function of their well-developed locomotor musculature may be to move between areas with a good supply of food and to leave when an area runs out of food, which lancelets obtain by attaching to the substrate and filtering fine particles from water currents driven through their gill slits. Permanently attached filter feeders need constant water currents. Lancelets shift feeding sites when necessary (Schomerus et al., 2008). They also swim to spawn. The possibility of locomotion over quite long distances is suggested by species of lancelets (*Asymmetron*) that specialize in feeding on particles around decaying whale carcasses on the deep-sea bottom (Kon et al., 2007). These are rare and scattered resources, and long-distance search seems certain to be advantageous

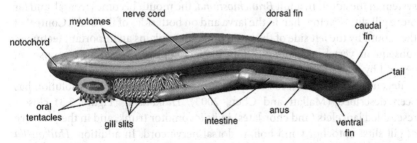

Figure 3.1 *Branchiostoma*, a lancelet. Note the asymmetrical position of the mouth on the left side, as in the larva but not the adult.

in such a way of life, which has (on genetic evidence dating the point of origin of *Asymmetron* species) been used since remains of giant marine reptiles, not whales, were the source of food.

Hints (Jefferies and Lewis, 1978) as to the way in which greater bodily mobility may have evolved in filter-feeding ancestors of chordates are provided by early echinoderms (Figure 3.2) and Cornutes, which moved by using a segmented tail (probably derived in evolution from an attachment stolon) to dig into the substrate and pull themselves along backwards. They fed on the sea floor, where they lay on their right side, with gill slits only on their left side.

Asymmetry was present in the line of evolution that eventually led to humans, and the evidence from Cornutes suggests that it too may have originated from lying on one side. Lancelet larvae are strikingly asymmetrical, with the mouth on the left side of the body. Unlike the symmetrical adults, which feed exclusively on small particles, the lancelet larva takes some prey items that are large relative to its mouth. It feeds for much of the time, whilst sinking slowly with the mouth open, and the left side facing downwards (Webb, 1975); this alternates with shorter periods of swimming upwards with mouth closed. Glandular secretion makes the lower lip and left side around the mouth sticky enough to attach prey (mainly copepods). Evidently the mouth muscles then dilate the mouth, allowing the prey to be engulfed, and then grasped by the muscular gill bars and ingested. The anterior mouth musculature is striated, allowing rapid contraction (Webb, 1969). Stokes (1997) notes that, when larvae land on their left side on the substrate (i.e. mouth down), they remain stationary for up to a minute, but they do not do so after landing on the right side; this suggests that they test the substrate for potential food items by the mouth area.

Left positioning of the mouth means the neural connections necessary for detection and seizure of prey are likely to be on the left side of the central nervous sysem. In fact on the left side of the larva's head there is a nerve plexus that innervates the mouth (Drach, 1948). The mouth dilator muscles disappear in the adult, their function being gone. However, Jefferies and Lewis (1978, p. 267) noted that the nerve plexus of the mouth cavity (presumably now solely sensory) is still connected only with the left side of the anterior central nervous system in the adult. In adult *Branchiostoma*, the mouth becomes frontal, and far more gill slits develop than in the larva and on both sides of the body. Control of the mouth by the left side of the brain probably explains an important feature of subsequent vertebrate lateralization, namely control of feeding responses in general by the left side of the brain, as discussed later.

Recently an important fossil representing the next step in our evolution has been described (Mallatt and Chen, 2003). *Haikouella* (Figure 3.3) closely resembled lancelets (and chordates) in its locomotor trunk, and in the presence of gill slits, notochord and hollow dorsal nerve cord. In addition, *Haikouella* had intrapharyngeal denticles that would have allowed larger food items, once engulfed, to be crushed by pharyngeal contraction, like that performed by

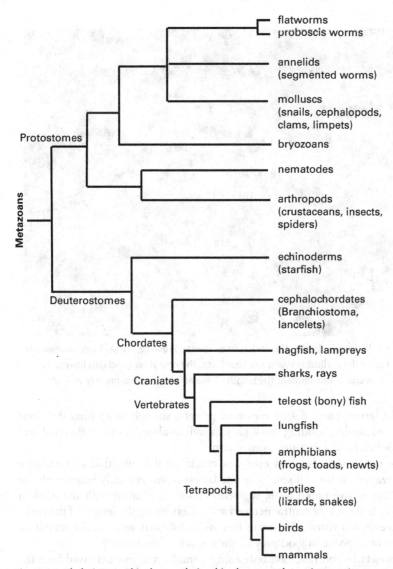

Figure 3.2 Phyletic tree. This shows relationships between the main organisms mentioned in the text. Also amongst the animals discussed in the text, *Caenorhabditis* is a nematode, and trilobites are ancient arthropods. Echinoderms include fossil forms such as Cornutes, which are discussed in the text as giving clues about how chordate asymmetry may have originated. The five-rayed body of familiar modern echinoderms like starfish or sea urchins was acquired mysteriously during evolution from ancestors with bilateral symmetry. The key form *Haikouella* has not been formally assigned to a phyletic group (and is not shown in the figure), but it fits in as directly ancestral to craniates. Note that the lines indicate connections between groups in evolution but they do not indicate any accurate time-line.

Figure 3.3 *Haikouella* fossils. This rock is from Yunnan Province, China. These animals are thought to be the earliest craniate-like chordates. The neural cord and gills have been indicated on one of the animals. The length of one animal is approximately 20 mm.

lancelet larvae. Large paired eyes were present, strongly suggesting that food items (no doubt including small prey as well as algae) could be detected and approached from a distance using vision.

The possession of paired eyes was crucial to the evolution of vertebrate lateralization, as we will see. Chordate lateralization probably began with the evolution of swimming using segmental muscles, coupled with the need to coordinate these by a central nervous system extending the length of the body. Paired eyes and visual selection of food items followed, and was dependent on control of response to food/prey by the left side of the brain.

Lancelets, by contrast, have only a single, small eye in front, derived from the apical organ (Lacalli, 1994). The paired eyes of vertebrates appear to have evolved from this. During development, the vertebrate eye rudiment is initially unpaired, and then divides into two (e.g. in the toad *Xenopus*; Li *et al.*, 1997), which suggests an evolutionary origin by division.

Although lancelets have many photoreceptors in ocelli within the central nervous system, these cannot have been involved in the origin of the paired eyes of vertebrates, since (unlike the frontal eye, which has retinal receptors) they contain the incorrect type of photoreceptor (Lacalli, 2008). The photoreceptors of lancelet ocelli use a photosensitive pigment, melanopsin, which is quite different from that used by retinal receptors involved in image-forming vision (Gomez et al., 2009). Melanopsin is the visual pigment used exclusively to sense circadian rhythms in light by cells in the retina and, in most vertebrates other than mammals, within the brain itself (Hazlerigg and Loudon, 2008). Apparently the neural structures that allow activity rhythms to be controlled by light have remained fundamentally the same throughout chordate evolution.

The larval lancelet eye is composed of four rows of cells, with a first row of six photoreceptors. During the day lancelet larvae usually hover in the light near the surface of the sea, and position themselves so that the eye is shadowed as much as possible from the sun (Stokes and Holland, 1995). As a result, the lateral photoreceptors are better illuminated than are the medial ones, and changes in illumination of left and right photoreceptors might allow detection of the movements of nearby objects (Lacalli, 1996). What might follow such detection remains to be established; sudden leaps might allow escape and/ or leaps at prey. In either case, increased reliance on laterally placed photo- receptors used in this way might have led to the evolution of paired eyes, via the establishment of left and right pits holding groups of receptors, so that left–right differences in illumination would be increased.

Paired giant cells (large paired neurons: LPN) within the 'primary motor centre' are of especial importance. The left giant cell receives direct and fast input from the frontal eye, whereas the right has its major input from the right 'tectum', and so is affected by sensory information that has already been pro- cessed to some extent and is presumably available more slowly. This includes input from sensory cells at the tip of the body, presumably providing tactile and/ or chemical information. Both LPN are linked to motor neurons, and have been suggested to control whether swimming patterns are slow or fast (Lacalli, 2002). Fast and slow swimming are mediated by different divisions of the trunk myotomes, under quite separate control, and no doubt with different functions. Slow swimming is probably used for vertical migration in the water column across day and night. The muscles involved in slow swimming are controlled by neurons that receive inputs from (amongst other inputs) the first dorsal ocellus within the central nervous system (Lacalli, 2002). Fast swimming would be appropriate for escape and/or for leaps at prey. Rapid response to light changes would almost certainly involve the left LPN and its fast input from the frontal eye.

### 3.1.1 Early vertebrates with two eyes

During evolution of the vertebrates, two eyes replaced the single eye and new visual mechanisms appeared. Fish-like swimming became possible, with the body stabilized rather than rotating in the direction of locomotion. Hence, it became possible for the evolving vertebrates to perform directed avoidance or approach on either side. The right eye assumed the role of detecting prey, and prey capturing remained under control of the left side of the brain.

An evolutionary stage in which visual input from paired frontal eyes was used in the selection of food items, and for the preparatory opening of the mouth, still on the left side of the head, may explain why the otherwise puzzling crossover of the vertebrate optic nerves (optic decussation) might have originated (Chapter 2). The right eye would receive input from prey movements, both when prey was ahead, and when it was to the right. In either case a turn to the right would tend to bring the mouth to bear on the prey; this would be initiated by right eye information, which would have to reach mouth control mechanisms in the left side of the brain to allow the mouth to be opened in order to seize the target. The use of right eye input to initiate and control prey capture survives in many living vertebrates (teleost fish, Miklósi and Andrew, 1999; Miklósi et al., 2001; teleosts and newts, Giljov et al., 2009; lungfish, Lippolis et al., 2009; birds, Andrew et al., 2000, Tommasi et al., 2000; toads, Vallortigara et al., 1998, Robins and Rogers, 2004; frogs, Robins and Rogers, 2006; and see Chapter 1).

Prey detected on the left would need little or no body turning and no particular specialization. The fast visual input from the frontal eye to motor structures on the left, already present in lancelet larvae, might in consequence have remained with the right eye, so that turns towards prey seen on the right could be initiated. As locomotion became more effective, turning to either side on the basis of visual input would have become essential both for prey catching and for avoidance of predators. This would have required the evolution, on both sides of the brain, of new structures controlling locomotor turning in either direction. However, apparently left brain structures remained responsible for the initiation of seizing prey with the mouth.

As mentioned above, evidence from lancelets suggests that the features needed for further evolution towards a vertebrate-like condition were predominantly those of the larva and not the adult. Neoteny (appearance of reproduction in larval stages, which sometimes occurs in lancelets today) may have been responsible for further evolution (Bone, 1972; Mallatt, 1985).

The conodonts are believed to have been fundamental to the evolution of vertebrates, because they were the first possessors of true vertebrate teeth (Dong et al., 2005). They had fish-like segmented trunk musculature, a notochord and a tail fin, which allowed free swimming (Figure 3.4). They appear to have used their large paired frontal eyes (which could have been moved by extrinsic eye muscles: Donoghue et al., 2000) in locating prey (Zhuralev, 2007). The prey were then grasped and crushed within the pharynx by the complex teeth (which were intrapharyngeal like the denticles of *Haikouella*).

As small free-swimming predators in the plankton, conodonts shared their niche with an extant group, the arrow worms (Chaetognaths), quite unrelated to the chordates. The resemblances between the two groups (Szaniawski, 2009) are made more striking by the fact that both evolved the ability to immobilize prey by venom. In the conodonts, this was injected along grooves in their teeth,

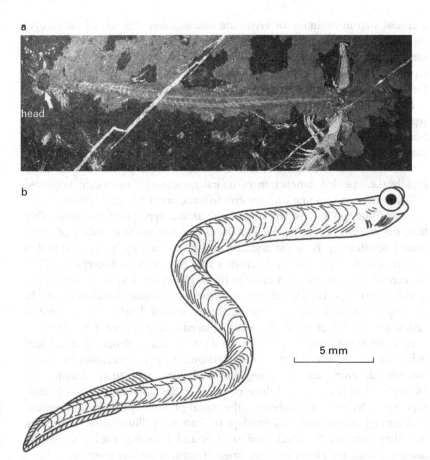

Figure 3.4 A conodont. a, Photograph of a fossil with head indicated and the body (40 mm long) spanning the image. b, A drawing representing a conodont. Early in evolution, conodonts were highly successful small marine predators. The fossil shows chordate credentials in its possession of a lancelet-like body, allowing free swimming, together with the crucial new possession of a pair of large eyes. Photograph and drawing supplied by Emeritus Professor Richard Aldridge, University of Leicester.

an ability that may explain their long survival without the evolution of jaws, despite competition from early fish.

Arrow worms and chordates differ in that the eyes of arrow worms, although paired, have limited numbers of photoreceptors (Berezinskaja and Malakhov, 1995), look upwards and are immobile. Non-visual senses, such as those detecting changes in hydrodynamics and chemical signals, seem to be crucial to the perception of prey (Rieger *et al.*, 2011). Hence, it is likely that, in organisms such as conodonts, early reliance on advanced optic inputs from the right and left was a crucial step in evolution of vertebrate lateralization. The use of paired eyes capable of independent scanning (and so of use for different purposes), but at the same time progressively requiring more and more integration into a single representation of the world, near and far, would have brought increasing need for the collaboration of differing specializations of right and left sides of the brain.

Early Cambrian forms of life appear to have added vertebral elements, and so represented the appearance of the vertebrates in evolution (Shu *et al.*, 2003). Further vertebrate features were added with the evolution of jawed fish. The dimensions of oculomotor nerves in placoderms (early, jawed fish) suggest the evolution of myelinated nerves, which can conduct action potentials much more rapidly (i.e. speeded conduction of neural messages). This meant improved ability to use fast scanning with the eyes to track prey (Zalc *et al.*, 2008).

In addition, photosensitive organs with retinal-type photoreceptors, other than the paired eyes, have played an important role in the evolution of vertebrate lateralization. Both the asymmetrical pineal and the parapineal bodies appear to be derivatives of photosensitive structures with such receptors within the central nervous system, similar to the large unpaired lamellar body of the lancelet larva; pineal and lamellar receptors are very similar (Lacalli *et al.*, 1994). In lampreys, teleost fish and reptiles, the parapineal develops from a rostral vesicle and the pineal from a posteriorly placed vesicle (Yáněz *et al.*, 1999).

In lower vertebrates the pineal generally responds to dimming of ambient light: as shown in tadpoles of the toad *Xenopus*, the pineal receptors fire briefly following dimming, and swimming is evoked, presumably to avoid a predator (Roberts, 1978; Jamieson and Roberts, 1999). In the ammocoete larva of lampreys (primitive jawless vertebrates), the pineal photoreceptors fire in response to dimming of light and rapidly adapt to changes in illumination, suggesting that they function to detect shadows (Pu and Dowling, 1981). As will be discussed later, the pineal and parapineal structures are asymmetrical.

## 3.2 Left hemisphere and sustained response

An important function of the left side of the brain and the right eye is to sustain a response once it has been initiated and to avoid distraction by stimuli that might evoke other responses. This, of course, is essential in pursuing and capturing prey. Two regions of the brain have specialized roles in such behaviour. They are the parapineal gland and its connected habenular nucleus on the left side of the brain. The habenulae are located in the epithalamus and, although there are habenular nuclei on both sides of the brain, in many species the left habenula is larger than the right (see also Chapter 1 and Figure 1.1), or it may even have more lobes on the left than on the right side.

In lower vertebrates the parapineal, which may retain an eye-like structure, typically supplies the left habenula, or is involved in early development of the left habenula before it itself disappears. Although there is substantial variation in parapineal anatomy, a recognizable parapineal, distinct from the pineal, is always present (Concha and Wilson, 2001). In teleosts the parapineal lies on the left side of the pineal recess, and it supplies the left habenula (Vigh-Teichmann et al., 1983). In the lamprey, there is a separate nucleus on the left, fed by the parapineal (Yáněz et al., 1999), which supplies both the left habenula and the left side of the major target of the left (and right) habenulae, the interpeduncular nucleus (IPN). The habenulae were originally asymmetric, and in lower vertebrates their asymmetry is still strongly involved in the balance between left brain and right brain functions.

The most elaborate form taken by the parapineal, the parietal eye of some lizards (Figure 3.5), is responsible for the use of a clock-compensated sun compass in orientation during long-distance travel (e.g. when returning home after displacement; Freake, 1999, 2001; Foa et al., 2009). The lizard parietal eye retains the primitive neural connection with the left habenula (Engbretson et al., 1981).

In zebrafish, both left and right habenulae have lateral and medial divisions, which correspond to similar divisions in mammals (Bianco and Wilson, 2009). The lateral division, which is much larger on the left, projects predominantly to the dorsal interpeduncular nucleus (dIPN). The medial division, which is larger on the right, projects mainly to the ventral interpeduncular nucleus (vIPN) (Aizawa et al., 2005). The dIPN projects to the midbrain central grey, where its input opposes the initiation of intense fear: inactivation of the left lateral habenula promotes freezing to a negative conditioned stimulus (Agetsuma

Figure 3.5  Parietal eye of a lizard. The location of the parietal eye is indicated.

*et al.*, 2010). This is consistent with the left side of the brain controlling pursuit and capture of prey, as already discussed; clearly avoidance has to be inhibited if pursuit is to be successful.

In mammals, the lateral habenulae supply the central grey, as in the zebrafish, as well as other structures (Sutherland, 1982). The central (periaqueductal) grey is involved in both the promotion of violent attacks (Tulogdi *et al.*, 2010) and the reduction of response to pain (Leith *et al.*, 2010). There is left brain involvement in the inhibition of intense emotion, including fear (discussed later in this chapter and in Chapter 5). In mammals, the habenulae themselves appear to be symmetrical and, therefore, functional asymmetries probably originate from higher centres of the brain. It is possible that early in evolution the parapineal and its associated ganglion were both involved in sustaining a course of pursuit of prey against interference by avoidance, whilst (based on evidence from lizards) using light from above to remain on course.

There is direct evidence for continuing habenular involvement in pursuit of prey in mammals. The mammalian lateral habenular nuclei contain neurons that fire during targeting head movements used in pursuit (Sharp *et al.*, 2006), implicating these nuclei in sustaining pursuit. This involvement of the lateral habenulae in rats in head movements made while chasing a target appears to act, at least in part, through the IPN (Clark *et al.*, 2009); interestingly, lesions of the IPN also impair navigation by distant features. The IPN projections to the hippocampus may be involved, allowing involvement of hippocampal spatial analysis (Clark and Taube, 2009).

Chemosensory inputs to the left side of the forebrain sometimes drive sustained travel (migration) in fish, quite separately from evocation of species-specific (emotional) responses via the right forebrain. Westin (1998) has shown that migration of eels is disturbed when the left nostril (which supplies olfactory regions in the left side of the brain) is blocked but not when the right is blocked: eels with the left nostril blocked migrate more slowly and make more incorrect choices at decision points. A special role of the left side of the brain appears to be in sustaining a chosen route or orientation using odours, as it is for visual pursuit in other species. In other words, the left side of the brain has a long evolutionary role in sustained pursuit of prey and following of a route of migration (Chapter 6 discusses this in migrating birds).

## 3.3 Right hemisphere and emotion

The right side of the brain is specialized to express intense emotions (Chapter 1). Species-specific behaviour, including sexual, fear and aggressive behaviour, is commonly lateralized to the right side of the brain. This lateralization began in fish or perhaps even earlier in the first vertebrates.

In the zebrafish, a particular population of mitral cells in the olfactory bulbs supplies only the right habenula (Miyasaki *et al.*, 2009). In view of the special role of the medial habenulae in evoking intense species-specific response to releasers, and of the fact that it is the right medial habenula that is enlarged in the zebrafish, it seems likely that this input is important in response to chemical signals (pheromones), evoking behaviour such as sexual or fear responses.

An overall pattern very widely distributed amongst vertebrates is use of the left eye (involving activation of the right side of the brain) to promote attack. This has been recorded in toads (Robins *et al.*, 1998; Vallortigara *et al.*, 1998), lizards (Deckel, 1995), domestic chicks (McKenzie *et al.*, 1998; Vallortigara *et al.*, 2001), baboons (Casperd and Dunbar, 1996; Drews, 1996) and horses (Austin and Rogers, 2012). In the chick, facilitation of sexual behaviour also occurs with left eye (right hemisphere) use (Rogers *et al.*, 1985). In humans also, facilitated evocation of sexual behaviour is associated with right hemisphere control (Tucker and Frederick, 1989). Right hemisphere-associated mania often involves exaggerated sexual behaviour, whereas right hemisphere stroke is more likely than left hemisphere stroke to depress sexual function. Furthermore, sexual arousal is accompanied by greater desynchronization of the central or posterior electroencephalogram (EEG) in the right than in the left hemisphere (Tucker and Frederick, 1989).

Fear behaviour is facilitated by activity in the right, but not the left, amygdala, as shown in the rat (Adamec and Morgan, 1994). In the chick, the homologue of the amygdala on the right side of the forebrain controls distress calling (Phillips and Youngren, 1986). Fear and escape responses are more readily released in the toad when a model predator is detected in the left monocular visual field than in the right monocular visual field (Lippolis *et al.*, 2002). In humans, horror films are more disturbing when seen in the left compared with the right visual field (Dimond *et al.*, 1976). Readiness to respond to powerful releasing stimuli (stimuli that trigger fixed patterns of behaviour, as in escape) could also underlie the preferred use of the left eye by larval zebrafish and other species to view their reflection in a mirror (Sovrano *et al.*, 1999; Andrew *et al.*, 2009a). Davidson (1995) reviews evidence showing that right hemisphere control results in facilitation of negative states (e.g. fear, disgust) in humans. Since right hemisphere control is also associated with more intense autonomic activity (Lane and Jennings, 1995), it has been argued that its function is associated with avoidance, whereas left hemisphere control is associated with approach (evidence of this in dogs was presented in Chapter 1; Quaranta *et al.*, 2007). This specialization of the right hemisphere extends to other sensory modalities: Ehrlichman (1986) showed that negative odours delivered to the right nostril are rated as more unpleasant than the same odours presented to the left nostril. Since there is no crossing the midline (decussation) of olfactory input, this is probably associated with right hemisphere assessment.

Not only unpleasant emotions and responses like fear are expressed more intensely when the right hemisphere is in control. Evocation of sexual behaviour due to such control (discussed previously) provides strong evidence in support of this idea. Left eye use is also more likely in a range of situations in which pleasant stimuli are being viewed. Lizards tend to turn right, to look with the left eye, when approaching a food reward (Bonati *et al.*, 2008). Mangabeys increasingly view food items with the left eye as palatability of the item increases (de Latude *et al.*, 2009). In humans, the use of the left eye, achieved by turning the head to the right, occurs both when an emotive picture is viewed and immediately after kissing (Ocklenburg and Güntürkün, 2009). Right hemisphere control is thus associated with emotion. This evidence suggests that the right hemisphere is used when any kind of intense emotion is expressed.

By contrast, when it is necessary to inspect an important and relevant stimulus, without responding to it prematurely (e.g. by fleeing), the right eye and left hemisphere tend to be used. This is the case in teleost fish for approach to view a predator (Bisazza *et al.*, 1997a, 1997b; Thomas *et al.*, 2008). The use of the right eye during approach allows distracting stimuli to be ignored, after brief inspection, even though they might be treated under other circumstances as potentially dangerous. Zebrafish view unfamiliar objects with the right eye on first encounter, but with the left eye thereafter (Miklósi *et al.*, 1998). Right eye use allows initial inspection of the stimulus by making avoidance less likely, followed on later encounters by use of the left eye in order to assess many properties of the object and thus to establish its degree of familiarity (discussed also in Chapter 5).

Comparable effects may occur in humans, since activation of the left frontal region of the cortex occurs in anger when the confrontation is about to be resolved by approach (Harmon-Jones *et al.*, 2010). In this case, increasing inhibition (by the left hemisphere) of powerful emotion is likely to be occurring.

In primates, individual differences in boldness or emotion sometimes correlate with hand preference, suggesting effects associated with overall bias towards control by the left or right sides of the brain. For example, right-handed chimpanzees and common marmosets approach and touch novel objects sooner than do left-handed chimpanzees or marmosets, respectively (Hopkins and Bennett, 1994; Cameron and Rogers, 1999). Left-handed marmosets prefer to explore visually rather than handling a novel object, and they also show persistent elevation of cortisol after return to the home cage, following a period in a strange cage, whereas this response is absent in right-handed marmosets (Rogers, 2009). Left-handed Geoffroy's marmosets are less likely to sniff at novel food than are right-handed ones, and they freeze for longer after hearing the call of a predator (Braccini and Caine, 2009). Such individual differences probably depend on which hemisphere is used following detection by the right hemisphere of emotion-provoking stimuli (discussed further in Chapter 5). In these examples, the hand preference was measured as that preferred to pick up pieces of food from the floor or a bowl.

This task does not require coordinated use of the hands and it is so simple that either hand can be used without difficulty (more details in Chapter 6).

In humans, Merckelbach *et al.* (1989) found no clear evidence of any elevation of anxiety in left-handers. The greater success of left-handed humans in competitive sports is commonly attributed to the fact that left-handed opponents are more rarely encountered (see Chapter 2); however, the incidence of left-handedness is considerably elevated in societies in which homicide is common, being present in the most extreme case in about a quarter of the population (Faurie and Raymond, 2005). Such an elevation suggests, as an alternative (or additional) explanation, that left-handers may be more likely to proceed to, and sustain, the most intense levels of response in conflict, and that this may be advantageous to them in some human societies (Billiard *et al.*, 2005).

## 3.4 Right hemisphere and spatial abilities

Amongst vertebrates, a separate set of abilities is widely associated with use of the left eye and right hemisphere. Regions of the right side of the brain are specialized to assess multiple properties of objects and of the environment. This function contributes to assessment of novelty and familiarity, as well as to assessment of spatial relationships between objects using global or geometric cues (see Chapter 2). For example, left eye use enhances the ability of chicks to use geometric cues to locate the centre of an arena, following distortion of its shape or a change in its size (Tommasi and Vallortigara, 2001) and to orient by distant cues (Rashid and Andrew, 1989). Chicks also respond to changes in spatial context when using the left eye and ignore them when using the right (Andrew, 1983, 1991a). Rats are able to use topographical cues to find the refuge platform in a Morris water maze when they use their left, but not their right, eye (Cowell *et al.*, 1997). Left eye use allows use of such distal features (Cowell *et al.*, 1997). Also in rats, lesions of the right but not the left parietal cortex impair the use of such distant features to control path direction taken in a Morris water maze (Adelstein and Crowne, 1991).

In humans, the right hemisphere has advantage in overall judgements of position (e.g. in estimating position by $X-Y$ coordinates: Kosslyn *et al.*, 1992). Moreover, left eye occlusion decreases accuracy in bisecting a line more than does right eye occlusion (McCourt *et al.*, 2001). Right hemisphere advantage in assessing faces (e.g. gender or expression) is also likely to be related to superior ability to assess multiple cues and the relations between them (Aljuhanay *et al.*, 2010).

In contrast, use of the right eye and left hemisphere is associated with attention to specific targets and memory of their properties. When returning to a site where they have stored food, after being away for a brief interval, marsh tits use topographical cues to guide their searching when using their left eye, and local cues when using their right eye (Clayton and Krebs, 1994b). Similarly, when the site of food is locatable both by a local cue and by distant cues, chicks using the right eye make more use of local cues than do chicks using the left eye (Tommasi and Vallortigara, 2001; and see Chapter 2).

Primates show preferential attention to global aspects of complex stimuli when using the left visual field, but to local aspects when using the right. This is demonstrated by matching tasks, in which a pattern (a large letter made up of smaller letters; e.g. a large 'H' made up of smaller 'A's) has to be matched against a stimulus letter. Presentation of this pattern to the right or left hemisphere was achieved by requiring the subject to fixate a central point while the letter was presented very briefly in the extreme right or left visual field. With presentation in the right visual field there was a preference to use the local cue (A), and with left visual field to use the global cue (H). This was found in humans and baboons in exactly the same test (Deruelle and Fagot, 1997) and in chimpanzees (Hopkins, 1997).

When both eyes are in use during training and testing (no eye-patches), differences between specializations of right and left sides of the brain are revealed by effects on counting. Chicks were trained to select a particular position in a row of targets (e.g. fourth or sixth in a series of 16) along which they walked. If before testing the row was rotated 90°, so that it lay ahead and perpendicular to the line of approach (i.e. the chick could go to the left and peck at positions four or six, or to the right and peck at positions four and six on that end of the row), chicks preferentially chose to count from the left end of the row (Rugani et al., 2010). The neural systems fed by the left eye thus took the lead in choice on an advanced feature of pattern analysis. This finding was supported by the fact that altering the spacing between the targets removed the bias to start from the left, although it permitted correct counting (Rugani et al., 2011).

How might the right hemisphere (using the left eye) have acquired these abilities during evolution? One possibility is suggested by the following: teleost fish impose standing biases on the Mauthner cells (on either side of the hindbrain) that control sudden escape, so that such escape is directed towards a refuge or exit route without any delay caused by carrying out more detailed visual investigation of the environment (Eaton and Emberley, 1991). The animal maintains a memory of the nearest escape route and this information is updated, as shown in frogs, each time the animal moves (Ingle and Hoff, 1990). Updating by the left eye (right hemisphere) would also be advantageous during periods of pursuing prey, when the systems fed by the right eye would be strongly attending to by the movements of the prey. At the same time as the right eye is attending to prey, systems fed by the left eye (right side of the brain) for sustaining and updating information about layout of the immediate environment would be important. This ability might already have evolved in a conodont-like ancestor.

## 3.5 Vision and hearing

Asymmetry of the visual system in vertebrates is well exemplified by that of birds. In birds, at least some asymmetries in vision arise from the large midbrain structures, the tectofugal system in pigeons and the thalamofugal system in chickens (described in Chapter 4). The main relay nuclei from the optic tecta to the forebrain are the rotundal nuclei (see Figure 4.2). In pigeons the rotundal nucleus on the left side of the midbrain receives almost equal input from the right and left optic tecta; consequently the left hemisphere receives input from both eyes, unlike the right hemisphere, which has input predominantly from the left eye (Güntürkün, 2002). Further, the left tectum is better able to inhibit its partner, so that, once inputs from the right eye become the focus of attention, such control tends to be sustained more readily than would be the case for inputs from the left eye (discussed also in Chapter 5). Hence, birds have clear asymmetry of their visual pathways and behaviour.

In birds, although convergence for binocular fixation and yoked eye movements do appear once a target has been identified, independent scanning by the two eyes is common (Wallman and Pettigrew, 1985; Martin, 2009). Each eye can also focus independently, which is important when using the lateral visual fields (Schaeffel et al., 1986). In the chick, collaboration between the two sides is not confined to periods of binocular convergence. Head saccades to bring this about have amplitudes of about 10° and are given in regular series, which can lead up to the final fixation position of the head or be displaced from this by 5° (Tommasi and Andrew, 2002). At angles as great as 25° or 35°, there is a preponderance of right eye use. This latter effect is presumably associated with use of the right eye system in controlling selection of, and response to, targets.

In mammals there has been a pronounced shift away from asymmetries of tectal function and anatomy, as described in birds. The superior colliculus (the homologue of the optic tectum) is greatly reduced in size in comparison to the equivalent in birds (i.e. relative to the size of the whole brain). The reduction in the role of the superior colliculus in mammals is likely to have been evolved with changes in the way senses were used. Great improvement in olfaction is suggested by fossil evidence as a key first step in the evolution of mammals (Rowe et al., 2011) and evidence shows that this too is lateralized (Chapter 5). Vision is likely to have been reduced in importance, if the assumptions that early mammals were nocturnal and lived in burrows are correct. Later, vision became very important to mammals and it evolved in ways that made it different from vision in birds and their predecessors. An important change in mammals was the much greater importance of conjugate eye movements, so that a single point of attention is a common feature of their vision. This step in evolution may have followed a period of greatly reduced reliance on vision.

Evolutionary pressures have also acted through changes in development. The appearance of eggs like those of birds, with a firm shell, allowed the embryo to take up a posture exposing only one eye to entering light. Exposure of the right eye (which normally looks outward at light entering through the shell) is necessary for the development of the normal left brain control of visually induced responses (details in Chapter 4). Lizard embryos (*Mabuia, Lacerta*) also take up a posture that potentially exposes the right, but not the left, eye to light (Pasteels, 1970).

Competition between right and left eye for establishing connections with brain structures may survive even in mammals. In rabbits the optic nerve from the right eye to the brain becomes myelinated before the one from the left eye, and the right eye opens first (Narang, 1977). Although it might be considered unlikely that effects of light input are important in this species since the young usually develop in burrows, some rabbits give birth in shallow nests that would allow exposure to light. As studied in rats and mice, retinal activity ('retinal waves') generate activity via the optic nerves, and this in turn drives activity in visual areas of the brain and affects the timings of establishment of connections within and between visual structures (Bonetti and Surace, 2010; Colonnese *et al.*, 2010). These timings and the competition between developing connections may well establish asymmetry within the visual system of mammals.

Auditory inputs reach both ears, in contrast to lateral visual inputs, which affect only one eye. Evidence from tests on humans shows that perceived position in space determines which side of the brain is involved in assessment of sounds, thereby yielding right ear advantage for verbal, and left for emotional content (Pierson *et al.*, 1983). This is true regardless of whether perceived position of the sound source is visually derived (e.g. position of a dummy loudspeaker), auditorily derived (sound position the only guide; Hugdahl, 1995) or caused by attention following instruction.

The possible preferential use by primates of the right or left ear according to the character of the sound, as shown by head turning, remains controversial (Teufel *et al.*, 2010) but evidence for right ear responsiveness in mice (i.e. attention to left hemispace) to motivating calls is considered in Chapter 5, and left and right side turning by dogs in response to different sounds is discussed in Chapter 6.

## 3.6 Evolution of language

The discovery (Nottebohm, 1970) that, in the canary, lesioning (destruction) of an important forebrain song area (hyperstriatum ventrale pars caudale or higher vocal centre, HVC) on the left, but not the right, led to major loss of song provided an enticing parallel to the evolution of human speech. Then, recognition of the absence of any detectible asymmetry in size or dendritic organization in this area (Nottebohm *et al.*, 1981) reduced interest in any such parallels. However, recent evidence has revived interest in comparisons between bird song and human speech (Zeigler and Marler, 2008). For example, a recent study has shown that Bengalese finches use syntactical information in processing of syllables and can even acquire artificial grammatical rules from synthesized strings of syllables (Abe and Watanabe, 2011).

The fact that the bird syrinx requires bilateral control (since it has paired sound sources) complicates comparison with glottal sounds generated by humans. Predominantly left-sided control of the syrinx (i.e. control from the left HVC), via the left nerve to the syrinx (Nottebohm, 1981), is normally present in canaries. However, in the next song season after left HVC lesion, a normal song is relearned, showing that the right HVC is capable of the same performance as its partner. In this species, at least, the brain retains plasticity.

It is now clear that left, right and bilateral control of the syrinx are sometimes all shown by the same species (Suthers, 1990; Wild et al., 2000). Cowbirds use the left side to produce the initial call cluster, and then shift to the right for later varied components (Allan and Suthers, 1994). In some species, right and left sides of the syrinx contribute different components of a single syllable simultaneously (Wild et al., 2000). At this level, therefore, it is interesting though difficult to make comparisons between lateralization of control of vocalization in birds and humans. However, the great ability of species such as parrots and corvids (crows, ravens, magpies) to copy patterns of sound (e.g. mimicry in Australian magpies; Kaplan, 2000) is important. Continuing neuroscientific and behaviour research examining the parallels and differences between learning of song by birds and learning of language by humans is fascinating and informative (Doupe and Kuhl, 2008).

### 3.6.1 Handedness, speech and language

Interest in lateralization began with the obvious presence of right-handedness in humans. It was assumed to be a uniquely human condition, underlying our ability to produce speech. This seemed to be confirmed by the finding that language is usually a function of the left hemisphere, thus explaining the use of the right hand in writing. Handedness was naturally used as a convenient index of the presence and direction of lateralization in humans, and later extended to animal studies. Unfortunately, a variety of different methods were used to measure handedness and this confounded interpretation of the results (see also Chapter 1).

Ideas about the evolution of handedness in primates were put on a new footing by the 'postural origins theory' of MacNeilage et al. (1987); see also MacNeilage (2007) for reconsideration and continued support of this theory. Left hand use in rapid strikes to seize small mobile prey evolved, it was argued, in insectivorous prosimian primates (early primates). It involved two specializations: (1) the right forelimb was used in support and so was specialized for sustained (prolonged) grasping, requiring strength, and (2) the left forelimb was used to strike at moving prey. MacNeilage (1998) specifically ascribed this postulated initial use of the left hand in primates to a pre-existing right hemisphere specialization for 'apprehension of the world'. Almost certainly this evolutionary step was associated with specialization of the right hemisphere for locating objects within the full visual panorama (i.e. with attention to both the left and right sides, compared with the left hemisphere's focus on the right side, as discussed further later in this chapter). In other words, the left hand (controlled by the right hemisphere) could readily capture insects on either side of an animal's midline.

Such left hand use has persisted throughout primate evolution. Squirrel monkeys use their left hand to strike at a difficult moving target (a fish), even though they show no clear hand preference in other tasks (King and Landau, 1993). A comparable specialization persists in humans (MacNeilage et al., 1987): even right-handers are more accurate in pointing at a suddenly presented small visual target when the left hand is used (Guiard et al., 1983). Although right-handed people show right hand superiority in accurate throwing, their left hand is superior in catching moving objects (Watson and Kimura, 1989; MacNeilage, 2007).

However, the key factor relating hand control to the evolution of language was the use of vision to control hand shaping for specialized grasping. The ancient role of left hemisphere structures in the control of sustained response would have been almost certain to result in a preference to use the right hand in grasping, once extensive and accurate hand shaping had evolved for different specific purposes.

The evolution of the visual control of hand postures and action can now be traced in the primate line of evolution (Rizzolatti et al., 1996; Rizzolatti and Arbib, 1998). Nerve cells in the brain known as 'mirror neurons' are found in the primate (macaque monkey) homologue of the area of Broca, essential for speech production in humans. These neurons fire when the test subject sees a specific action (usually of the hand, but occasionally of the mouth) performed by another, as well as during performance of the action itself by the monkey. In humans, there is some overlap of control of mouth and hand: the width of opening required for the one affects the other (Gentilucci et al., 2001). The representation of hand and fingers in the human area of Broca (left hemisphere) remains very like that in a corresponding area in macaques (Binkofski and Buccino, 2004). In humans, differences between the area of Broca (Figure 3.6) and its equivalent area in the right hemisphere correspond well with general right–left specializations: the area on the left is active during mental imagery of one's own limb, while that on the right is associated with position of the limb in extra-personal space (Binkofski and Buccino, 2004). The first is consistent with the control of response and the latter with assessment of the organization of the environment.

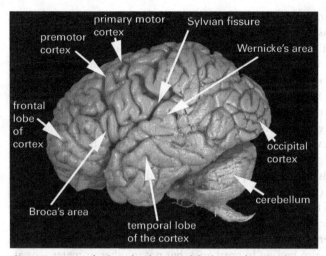

Figure 3.6 Human brain. Left side view of the human brain with areas mentioned in the text labelled. Note that the large area is the left hemisphere, and the cortex is the convoluted (folded) structure over the entire hemisphere. The smaller area to the right is the cerebellum, discussed in Chapter 5. The areas of the cortex concerned with speech and language are Broca's area and Wernicke's area. The Sylvian fissure is the large groove with the arrow pointing to its upper end.

Precise control of hand use became important in the great apes to access sources of food, such as extracting termites from holes in their nests (chimpanzee: Whiten *et al.*, 2001) or folding nettles into wads with the main stinging hairs facing inside before eating them (gorilla: Tennie *et al.*, 2008). Termite fishing in the wild actually has a bias to left hand use in chimpanzees (Lonsdorf and Hopkins, 2005; Hopkins *et al.*, 2009). It is possible that this left bias reflects both the type of task (e.g. precise positioning under field conditions of an irregular flexible tool) and the necessity of prior instruction in the wild: elaborate preparation of tools has to be learned (Whiten, 2005). However, the preferential use by chimpanzees of a finger of the right hand for extracting food from a tube held in the left (Hopkins, 1995) is quite unambiguous. The same is true for gorillas and bonobos tested on the same task, but not for orangutans, which instead have a left hand bias (Hopkins *et al.*, 2011). It is not clear why orang-utans are an exception, but they also show a left hand preference in fine hand and finger movements to manipulate parts of their face, as in cleaning their ears and teeth (Rogers and Kaplan, 1996). For reaching during feeding in the wild they use the right hand (Peters and Rogers, 2007).

The ability to place the thumb opposite each of the fingers on the same hand, especially the fifth digit, is considered to be a major step in the evolution of humans. It allowed grasping and manipulation of objects with the fingers and thumb. Thumb opposability appeared in the hominid line well before stone tools were used, probably as part of increased use of other more perishable tools that left no permanent record. Almécija *et al.* (2010) note that the early hominid form *Orrorin* (*Ardipithecus*), living about 6 million years ago, already had a human-like thumb and suggest that this may have been related to manipulation and preparation of food, as well as to the evolution of bipedality. Since the hand and foot, to some extent, share developmental control systems, enlargement and strengthening of the big toe that took place with the evolution of bipedal locomotion would have facilitated thumb enlargement for object manipulation. Right hand use in the preparation of stone tools was already present in the Oldowan culture around 2.6 million years ago (Whiten *et al.*, 2009).

Key evidence as to the course of evolution of hand control is provided by the external asymmetry of the brain: in humans the right frontal and left posterior poles of the two cerebral hemispheres are larger. This asymmetry, referred to as cerebral torque, is also present in great apes, but not in other primates – or at least not in rhesus monkeys (Pilcher *et al.*, 2001; Hopkins and Nir, 2010). The left occipital enlargement is also clear in chimpanzees, bonobos and gorillas (Balzeau and Gilissen, 2010). It is therefore likely to be associated, not with advanced vocal control, but with the appearance of superior ability to coordinate hand shaping. Handedness in chimpanzees specifically correlates in direction with asymmetry in the dorsal primary motor areas of the cortex involved in hand and finger control (Hopkins and Cantalupo, 2004), whereas there is no difference in direction of asymmetry between right- and left-handed animals in areas homologous to human language areas.

There are possibly precursors to these asymmetries in non-ape primates. For example, right hand preference in marmosets is correlated with length of the right lateral sulcus, as well as with thickness of the secondary somatosensory cortex (Gorrie *et al.*, 2008). The lateral sulcus is the equivalent of the Sylvian fissure in humans (Figure 3.6): in humans it is usually longer on the left side owing to the presence of Wernicke's area, which is essential for language (Figure 3.6).

There is an intimate association of mechanisms controlling hand shaping and speech in humans (Corballis, 2002; Lavrysen *et al.*, 2008; Corballis *et al.*, 2012); only right-handers are discussed here to simplify presentation. Right hand use, with no involvement of speech, involves more activation of the left occipital region of the cortex, including visual areas (Figure 3.6). Activation of the latter is presumably involved in visual control of the hand. Left hand use involves greater frontal activation, in particular of the right inferior frontal cortex, suggesting greater need for planning and executing actions by the hemisphere less specialized for such a purpose. The right premotor cortex is responsible for the control of complex coordination in bimanual tasks (van den Berg *et al.*, 2010), with a special role in preventing 'unwanted mirror activity' (i.e. performing with the left hand the action being currently ordered for the right hand by the left hemisphere).

Overall, it is therefore very likely that these asymmetries in hand control arose because of the acquisition of high ability to use the two hands in precise and complex manipulations, initially for access to new food sources. The hand of the recently discovered fossil *Australopithecus sediba* (extant about 2 million years ago) suggests increased manipulative ability, which perhaps preceded any obvious human-like enlargement of Broca's area (Carlson *et al.*, 2011); at the same time there are hints of frontal lobe changes, perhaps related to decision-making.

Discussions of the origin of human language have tried to decide between a gestural and a vocal origin. Both have a long history. One of the main ideas was that the left hemisphere's ability for skilled manual actions (of the right hand) evolved into sign language and later into speech (MacNeilage *et al.*, 2009). Another possibility discussed was that the left hemisphere's capacity to control manual action extended to control of the vocal apparatus. MacNeilage (2008) has highlighted the vertebrate-wide role of the left hemisphere in control of routine behaviour, as opposed to the role of the right hemisphere in emergency responses, and postulated that this led to its role in social communication and eventually human speech.

Modern discussion of the gestural origins of speech began with Hewes (1976). It is likely that gestures played a much more important part in communication by early humans than they do now in modern humans; the involvement of the area of Broca in the control of hand posture (Binkofski and Buccino, 2004), as well as in control of the motor sequences of spoken language, is important evidence of such an earlier stage in evolution. We have said that the homologue of Broca's area in macaque monkeys contains 'mirror' neurons, active during both the shaping of the hand and seeing such shaping by another; in humans, the posterior part of Broca's area is active both in language tasks and tool using (Higuchia *et al.*, 2009).

Hand representation in the chimpanzee primary motor cortex already shows asymmetries in synaptic density that correlate with hand preference in bimanual tasks (Sherwood *et al.*, 2010). However, there is no comparable asymmetry at the population level in a partial homologue of Broca's area (area 45) in chimpanzees (Spocter *et al.*, 2010), nor any correlation of such asymmetry with handedness in individuals (Schenker *et al.*, 2010). The evolution of vocal language is thus specifically associated in human evolution with left hemisphere changes in structures like Broca's area, and this has apparently not evolved in chimpanzees (or only at levels difficult to detect).

Left brain control of responding to motivating sounds has a longer evolutionary trajectory: it has been found in mammals other than primates, and also in birds. For example, lactating female mice search for their pups if they have been taken from the nest and are vocalizing. They do so provided that they can use both ears or the left ear is blocked, and not when the right ear is blocked

(Ehret, 1987). In other words, the right ear and left side of the brain are used. Functional asymmetries of auditory cortex are involved (Geissler and Ehret, 2004). One higher-order auditory field (which is in fact multimodal), the dorsal field, is activated only in the left hemisphere when the female hears and approaches pup calls. This may reflect the left hemisphere's ability to maintain attention to the perceptual properties of the target.

Domestic chicks turn the right ear to the sound of clucks, produced by the hen, when these are first heard, even when they are tested in darkness (Miklósi et al., 1996). A shift to using the left ear takes place gradually as the clucks become familiar. However, this left ear use is replaced by right ear use just before the chick approaches a familiar cluck presented with the addition of an unfamiliar flute sound (Andrew and Watkins, 2002). When two different clucks are played alternatively on the chick's left or right side, the chicks choose the one on the right when it is the first sound heard, otherwise they make no clear choice (Watkins, 1999). Again, the left brain appears to be responsible for maintaining the record of the target of response.

These findings for mouse and chick agree well with the evidence from rhesus macaques. In rhesus macaques, auditory cortical areas show greater activity in the right hemisphere in response to sounds, with the exception of species-specific calls (Poremba and Mishkin, 2007). Cutting the commissures connecting the areas involved abolishes this asymmetry. The authors argue that such commissurotomy prevents the left hemisphere from inhibiting activity of the right hemisphere. This effectively increases left hemisphere control of responses evoked by the sound, much as does use of the right ear by the mouse and the chick. In all three cases assessment of a call begins with use of the right hemisphere and is followed by use of the left hemisphere.

A human structure central to speech is Wernicke's area, which, like the area of Broca, is a specialization of the left hemisphere (Figure 3.6). It lies within the planum temporale (PT), which is larger on the left in both chimpanzees and humans (Hopkins and Nir, 2010), and in humans is largely made up of secondary auditory cortex, concerned with auditory discrimination and speech comprehension (Shapleski et al., 1999). At least two auditory areas especially involved with perception of vocalizations have been identified in primates. Rhesus macaques have a right hemisphere area sensitive to species-specific vocalizations, which is quite different from Wernicke's area (Petkov et al., 2008). A component of the human Wernicke's area (in the planum temporale) can be identified in chimpanzees (Chapter 5), and it is larger on the left (Spocter et al., 2010). This region on both the right and left sides is involved in processing species-specific vocalizations in both chimpanzees and other Old World primates (Spocter et al., 2010).

At least one reason for the leftward bias in the planum temporale, shown by the chimpanzee but apparently absent in monkeys (Hopkins and Nir, 2010), is likely to be the fact that there is learning of features of the 'pant hoot' choruses, which are used during encounters with neighbouring groups (Marshall et al.,

1999; Crockford *et al.*, 2004). During such inter-group calling, the vocal features that distinguish the callers from neighbours become more distinct than they do in solitary calling; decisions about what responses to make would inevitably dominate at such a time, and might promote left hemisphere control. Strikingly, both hand shaping and control of learned vocalizations show beginnings of the left hemisphere control, as seen in humans. The use of specific 'attention-getting' vocalizations by chimpanzees is accompanied by increased activity in the homologue of Broca's area, and the same is true of banging or clapping with the hands (Taglialatela *et al.*, 2011).

The incorporation of vocalizations into gestural communication is thus likely to have been influential from very early on in human evolution. Ability to learn sounds from others would have facilitated this. Importantly, chimpanzees use sounds, as well as gestures, to communicate with humans when begging for food (Hostetter *et al.*, 2001). Sounds made with the lips ('raspberry') are sometimes given in frustration or stress; such raspberries were incorporated into pant hoots by one individual, and this was then learned by others (Marshall *et al.*, 1999).

Since chimpanzees use their right hand to beg for food from both humans and chimpanzees, this bias is unlikely to be a specific bias imposed by humans (Meguerditchian *et al.*, 2010b). Left hemisphere control was thus present early in gestural communication. When the human faces away, chimpanzees are more likely to vocalize; crucially, the sounds made are not species-specific calls (food calls, grunts), but idiosyncratic sounds apparently acquired individually. When the human faces the chimpanzee, gestures are used, such as begging with the hand, and pointing at the food. Pointing at the food is associated with gazing alternately at the human and the food, and chimpanzees that vocalized while gesturing are more likely to gesture with the right hand (Leavens *et al.*, 2004). In the wild, begging for meat includes reaching towards the carcass, and covering the possessor's mouth with one or both hands (Gilby, 2006). A variety of different gestures, some with a clear relation to an overt attempt to obtain food, are thus used in communication by chimpanzees.

In chimpanzees there is rightward bias of the mouth when they produce such attention-getting sounds and a bias to the left for species-specific sounds. This parallels humans: here the right side of the mouth moves more in producing words, but the left for emotional sounds (Reynolds Losin *et al.*, 2008); remember that the left hemisphere controls right motor structures.

Baboons are of interest here, because they have evolved human-like sounds, which allow mouth and tongue movements to modulate resonances of the upper vocal tract (like human 'formants'), and so make more conspicuous friendly and greeting displays (Andrew, 1976). It is not known whether baboons show any more learning of their vocal sounds than other non-human primates, nor indeed whether infant baboons match their mother's facial gestures (as in rhesus macaques: discussed further below).

However, once vocal sounds with formants evolved in human ancestors, matching facial gestures by infants would have promoted the ability to copy such sounds.

Neonatal chimpanzees are capable of matching facial gestures (mouth opening, tongue protrusion) performed by friendly humans who have reared them, for a period comparable (about 8 weeks of life) with the corresponding period in human infants (Myowa-Yamakoshi *et al.*, 2004). In both species, the end of this period coincides with much greater interest in social interaction and a shift to a single facial response (smiling in humans and mouth opening in chimpanzees) used in social greeting. Infant rhesus macaques imitate lip smacking (a greeting display derived from grooming movements) and tongue protrusion (Ferrari *et al.*, 2009), but for a very much briefer period of life (first week only). The sight of such species-specific displays almost certainly evokes the same behaviour in response in all three cases, as is usual in displays. Clearly the special occurrence of such behaviour in infants has the function of promoting bonding between mother and child (as remains powerful in modern humans). In human evolution it probably took on a new importance in promoting the learning of vocalizations by infants, and promoting understanding of their use to influence the behaviour of others.

Hand gestures have the additional, special property that, when they became used in communication, they already had clear and specific purposes. The meaning of each would be reasonably obvious, and their potential supply is open-ended. Crucially, there is also potential enrichment of cognition, with frequent use in new contexts of concepts based on proposed actions. The addition of vocal signals would allow the identification of individuals, and membership of different social groups from immediate family to extended groupings, as is essential for proper communication. Such identification would be much more difficult by gesture alone.

### 3.6.2 Learning of vocalizations

A study by Mundinger (1970) was groundbreaking for the understanding of learning of vocalization in birds. In American goldfinches (*Carduelis tristis*) flight calls, which are often individually specific, are used by the female to recognize the male as he flies towards the nest to feed her. The members of a pair share at least one flight call, which may have been learned by the male from the female, or *vice versa*. Groups of finches (even if entirely male and including different species) living together imitate each other, producing flock-characteristic calls. The use of learned calls to allow membership of a group is widespread in birds (for a review see Hopp *et al.*, 2001). In one species of parrot there are individually specific calls, which identify family members, and seem to be used to specifically 'address' absent members (Wanker *et al.*, 2005). Comparable use of individually specific calls in communicating with absent members also occurs in dolphins (Sayigh *et al.*, 2007). Spear-nosed bats learn group-specific calls to allow membership of foraging groups (Boughman, 1998).

In avian species in which pairs form the basis of larger social groupings, there are typically male and female calls in Northern Hemisphere species but this is not so common in birds of the neotropics or Southern Hemisphere (Kaplan, 2008). In hill mynahs, exchanges between pair members involve stereotyped calls and answers; the replying partner has to use a particular call, but this must be different from the one just heard, since it must be appropriate to his or her sex (Bertram, 1970). Ravens also have calls that are characteristic overall of one or other sex, but here a bird may also come to use a call of its partner, which is therefore a call of the 'wrong' sex (Enggist-Dueblin and Pfister, 2002).

In the swamp sparrow, auditory units fire both to the sound of a particular note sequence and when the same sequence is performed by the bird (Prather *et al.*, 2008). The resemblance to mirror neurons in primates that fire both to the sight of a hand gesture, and when the gesture is performed, is clear.

## 3.7 Social structure and the origin of language

The ability to use signals deliberately to manipulate the behaviour of others to achieve a goal is probably widespread in vertebrates, although evidence for this (rather than for the emission of displays under motivational control) has not been widely sought. Nevertheless, play in dogs is an excellent example (Horowitz, 2009). Specific signals of intention to play, such as the 'play bow' (forequarters lowered and rear elevated) and open-mouthed play face have long been recognized. Attention-getting signals (e.g. bark, exaggerated retreat) are given to signs of inattention in the partner. Once attention is achieved, play signals become intense.

The results of attempts to teach chimpanzees and parrots human language show how far cognitive abilities necessary for language can progress before language itself evolves. Chimpanzees reared like human infants and taught American Sign Language are able to answer correctly and readily who, what and where questions, which identify very wide ranges of categories and properties (Gardner *et al.*, 1992). The insurmountable obstacles to the spontaneous acquisition of language-like communication by chimpanzees are clearly the difficulties of producing a vocal signal with specific meaning, and then of using it to convey information to attain a desired change in the behaviour of a conspecific. Grey parrots have been taught to use words to name objects and properties (Pepperberg, 1994); again, training had to involve conversation, in which humans asked each other similar questions about objects and were praised or corrected as required according to the reply given.

Increased need for accurate and detailed communication associated with organizing and planning group activities, such as hunting and gathering food, must have acted to promote both brain and body changes needed for advanced language. A key example of the latter changes is that, at some stage during human evolution, laryngeal descent (or rather the production of a vocal tract with a bend that is L-shaped, allowing the tongue to produce a constriction that can be varied) made possible the production of vowels very distinct in formant structure (cf. baboons, above). Recently evidence from primate vocal sacs has given the first solid dating of hominine changes related to speech (work by de Boer reported by Harvey, 2011). Such vocal sacs affect resonances below about 2 kHz, causing them to cluster together and reducing the ability to produce distinctive speech sounds. The presence of a sac can be detected from the structure of the hyoid bone; *Australopithecus afarensis* (approximately 4 to 2 million years ago) had sacs, but not *Homo heidelbergensis* (600 to 250 thousand years ago).

Why then has it, so far, been possible to teach parrots to communicate by speech, but not chimpanzees, even though chimpanzees have been taught to use sign language? The answer probably lies in consequences of human social structure. In birds, a social structure based on association of long-term pairs, with offspring, has promoted the evolution of vocal identifiers of pair, family and larger social groupings. Current human societies, quite unlike those of other apes, also show relatively long-term pairing, with men as well as women providing resources for rearing children. In fact, humans have a number of anatomical and physiological differences from other apes, which show that there has been protracted adaptation to this condition. In chimpanzees (and other primates with multi-male societies), oestrus is advertised by exaggerated sexual swellings (Nunn, 1999). These give information about the probability of ovulation and, since the dominant male guards the female only at the peak of receptivity, other males mate with her before and after this time. The swellings are not present in gorillas, as would be predicted from their social structure of a single dominant male with several females, or in orang-utans. In humans a similar absence of sexual swelling could have resulted from long-term pair formation; once this was acquired, advertisement of oestrus would tend to be disruptive of the pair. A hint that such change may have already begun soon after the divergence of humans from chimpanzees, which occurred at about 4 to 5.5 million years ago, is indicated by acquisition of pubic lice by the human line of evolution at about 3 million years ago (Weiss, 2009). The lice evolved in the coarse body hair of gorillas. Humans do not have such coarse hair except in the pubic region. Therefore, the appearance in humans of pubic lice suggests the evolution of pubic hair as a continuously present signal of sexual maturity.

Humans also evolved novel elaborations of social structure, promoting elaborations of communication. In general, humans appear to have evolved with strong selection at the level of the extended family, maximizing the chance of rapid production of a large grouping with a variety of social bonds (e.g. to sibs; Hill *et al.*, 2011), promoting successful competition with other such groups. Lateralization of language to the left hemisphere may thus have co-evolved with a number of stages in human evolution: vocal mimicry serving to generate signals identifying individuals and groups, and hand gestures communicating specific intended actions. The latter, at least, humans probably shared with chimpanzees. Both humans and chimpanzees can organize hand postures and movements as a unit, as shown by the usual compression in time, when two similar hand movements are repeated, so that they take little more time to perform than a single such movement. Such compression is absent in baboons (Ott et al., 1994).

As hand gestures, with their inherent property of clear reference to specific intentions and means to achieve these, became central to communication, the area of Broca (in the left hemisphere) inevitably elaborated. At the same time, there would have been a problem of identifying individuals and groupings absent from the communicating group. The evolution of vocal identifiers would have provided an ideal solution. This together with the need to collaborate in cooperative ventures such as group defence and hunting or food gathering would strongly select for advances in communication associated with changes in brain lateralization.

## 3.8 Lateralization in animals other than chordates

As discussed in Chapter 1, differences in the specializations of the left and right sides of the central nervous system and behaviour are not limited to chordates (reviewed by Frasnelli et al., 2012). This strengthens the conclusion that lateralization gives substantial advantages, since it has persisted, or evolved many times, in such diverse groups of animals. Animals relevant here fall into three major groupings: deuterostomes (vertebrates and other chordates and their echinoderm relatives), ecdysozoa (animals which grow by moulting their cuticular exoskeleton: e.g. insects and other arthropods) and molluscs (limpets, snails, cuttlefish, oysters, etc.). It is unlikely that groups other than deuterostomes had any equivalent of asymmetric control of use of the mouth, as described earlier (Figure 3.2).

It is probable that the common ancestor of metazoan animals specified the left–right axis (Vandenberg and Levin, 2009). Since this is also true of single-cell organisms such as ciliates, the same basic genetic mechanisms of specification of the left–right axis were probably present in the common ancestor of multicellular animals. Both midline and distance from midline must also have been specified.

It is unclear how much more than this was retained from a common ancestor. There is a striking resemblance (Tomer et al., 2010) between the genetic specification of the anterior-most central nervous system, with highest processing functions, in annelids (mushroom bodies) and vertebrates (pallium). The same seven genes are involved in both groups; some but not all are similarly involved in insects (*Drosophila*). The annelids ('worms') appear to be somewhat separate from the other groups discussed here, suggesting that shared features are retained from a remote common ancestor. In fact, genes that now specify properties of the front of the central nervous system in all three groups have probably been retained from a common ancestor without a central nervous system.

The most striking evidence that the left–right axis may have been specified very early in metazoan evolution is the involvement of orthologues (homologous gene sequences in different species) of the Nodal family in left–right bodily asymmetry in snails as well as in chordates (Boorman and Shimeld, 2002; Grande and Patel, 2009). However, this does not establish the point at which chordate asymmetries of the central nervous system appeared. The best candidates for a common ancestor of our group (deuterostomes) are the worm-like hemichordates. These express the neural patterning genes in the same anteroposterior way as in chordates, but lack a central nervous system (Lowe *et al.*, 2003). Instead it is the epidermal nerve net that is affected by the genes. Since hemichordates have minimal bodily asymmetry (an excretory pore only on the left), it seems likely that genetic specification of the left–right axis was present before central nervous system asymmetry was evolved.

The same genes (*Distal-less*, and its partners) are expressed in arthropod limb outgrowths, and in fish and tetrapod limb buds (Panganiban *et al.*, 1997). Some ecdysozoans also have such gene expression in the central nervous system. It is thus possible that a common ancestor of (at least) deuterostomes and ecdysozoans had homologous neural structures, perhaps sensory in function, which became associated with body outgrowths. Perhaps these served initially only to carry sense organs and then became responsible for locomotion. However, this does not necessarily mean that any asymmetries present in extant forms were derived from a common ancestor, which at best was a small unsegmented worm-like form (Baguñà *et al.*, 2008).

Amongst molluscs, gastropods (e.g. snails) have marked asymmetries of the central nervous system, in part at least because their bodily torsion is so marked, as a consequence of shell torsion. Largely speaking, slugs have external symmetry and symmetrical paired sensory tentacles, although some asymmetry is evident as in *Triboniophorus* (Figure 3.7), and asymmetry can be manifested in their behaviour. *Limax* slugs, trained to avoid a particular food odour, may hold the memory in either the right or left procerebral division of the brain (Matsuo *et al.*, 2010): lesions of one or other of these divisions show that this is true for at least a week. There is equal likelihood of storage on right or left, although right ablation affects memory more fully, hinting at some asymmetry, perhaps in learning. In the familiar snail *Helix* (which of course has obvious torsion), again the same tentacles must be used during testing as in training, if memory is to be shown (Friedrich and Teyke, 1998); in this case function was confined to one

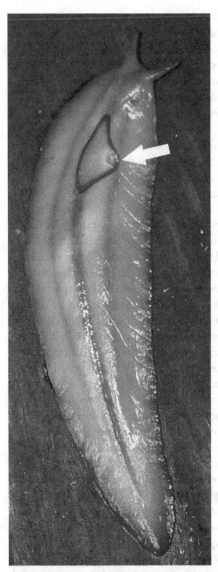

Figure 3.7 The air-breathing slug *Triboniophorus graeffei*, common in parts of Australia, showing some asymmetry of structure. The view is from above showing the slug's dorsal surface. Note the pneumostome to the right side of the midline (indicated by the arrow). This direction of asymmetry was present in all animals (total of 20) located in the nearby area. The triangular marking (red in colour) is also asymmetrical with its base on the midline and apex to the right.

side by manipulation of the chemosensory epithelium, so that brain structures on both sides were available and functional during any memory formation processes, although sensory input was confined to sense organs of one side. Hence, there is relative right–left independence but as yet no evidence of any right–left differences.

Cephalopods are the sister group to most other molluscs, with very independent evolution of a complex central nervous system (Kocot et al., 2011). Octopus and squid both combine behavioural asymmetry at the individual level with overall symmetry of their body. Individual octopuses have significant eye preferences for viewing a crab held outside the tank, but there is no population-level bias (Byrne et al., 2002, 2004). However, Alves et al. (2009) showed a population-level asymmetry in Sepia in turning bias across a series of unrewarded T-maze trials (see also Chapter 1).

Molluscs have evolved adequate eyes several times over, and even within the group to which Octopus belongs (cephalopods) there are species with large, efficient paired eyes, as well as the primitive Nautilus, which has a pinhole eye with poor resolution (Land and Fernald, 1992). Octopus readily learns visually to avoid attack, when this has been punished. The vertical lobe of the brain is necessary for such learning, and has resemblances to the hippocampus in that the main input consists of fibres running parallel and each making en passant contacts with many vertical lobe neurons. Transection of this input slows the acquisition of short-term memory of punishment (Shomrat et al., 2008). It is likely that many temporarily learned changes are mediated in Octopus by the nervous system of the arms, which contain about two-thirds of the neurons (Hochner et al., 2006), raising the possibility of comparison with the ganglia of arthropod appendages. Octopus is able to learn to control an arm to seize prey, using visual input to choose the tube into which the arm is to be inserted (Gutnick et al., 2011). It remains to be seen whether there is any preferred arm for such a task, and how information from such multiple effectors may be integrated into memory records.

A deep-sea squid (Histioteuthis: Wentworth and Muntz, 1989) does show population-level asymmetries. The left eye and left optic lobes are considerably larger than their equivalents on the right side (Figure 3.8). The left eye appears to be used to look upwards into the better-lit upper waters, possibly to detect predators. The smaller right eye looks downwards, perhaps searching for bioluminescence, probably of prey. Male squid (Sepioteuthis: Messenger, 2001) can give courtship colour displays to a female on one side, while giving a threat display to a male on the other. There is thus the capacity for considerable independence of motivational control on the two sides of the central nervous system; it remains to be seen whether this reflects any consistent behavioural lateralization.

Lateralization has been described quite widely in the Arthropoda, which is clearly a monophyletic group based on genomic analyses (Telford et al., 2008; Rota-Stabelli et al., 2010). Asymmetry of the body as a whole appears not to have been important in arthropod evolution; instead, asymmetry of appendages and of sensory structures seems to have evolved repeatedly, sometimes at very

Figure 3.8 *Histioteuthis*, a deep-sea squid. Note the asymmetry of the eyes. The larger left eye (on the right side of the photograph) looks upwards and the smaller right eye looks downwards. See text for details. Image modified from http://copperwitch.blogspot.com. au/2010/11/histioteuthis.html.

late stages in evolution. Arthropods lack the optic decussation that was so fundamental to chordate evolution: each of the paired eyes (like an appendage) feeds into the side of the central nervous system on which it lies.

From the start of arthropod evolution, interaction with the world was chiefly through appendages, use of which must have been asymmetrical in some cases. Asymmetric use of appendages in investigation and manipulation is present in both spiders (Arachnida) and the separate grouping of insects and crustacea with which trilobites are also associated. The spider *Scytodes globosa* uses the left first leg more than the right in both investigating and in capturing (silk wrapping) prey (Ades and Ramirez, 2002; see also Chapter 1). A wide distribution of such left leg use in assessment amongst arachnids (spiders and harvest men) is suggested by greater overall incidence of left leg damage in many species (Heuts and Brunt, 2005). A comparable asymmetry may be present in ants (Heuts and Brunt, 2005): a number of species move in two parallel 'streets' in their foraging routes. Both streets keep fellow ants in the other street on their left. Ants display another asymmetry also: a species that exchanges food from one ant to another shows a right antenna bias in the receiver ant (Frasnelli *et al.*, 2012). Significant bias towards leftward turning occurs in T-mazes in the giant water bug (Kight *et al.*, 2008).

In addition, asymmetry in healed injuries in very early Arthropods (trilobites of the Cambrian era about half a billion years ago) may stem from a comparable asymmetry in detection using the legs. Healed injuries (probably bites from a large predator) are more common on the right side (Babcock and Robison, 1989). One interpretation is that attacks by a predator from the left were more readily detected by the prey, allowing escape rather than death and consumption. This agrees with the other evidence just summarized. The other interpretation, which is that suggested by Babcock and Robison, is that the predators preferred to attack from the right; if true, this too might indicate that detection of prey was more likely with attack from the left, and that the predators may have learned this. Trilobites were in general feeders on the substrate, while some species were capable of taking larger items that needed crushing (Budil *et al.*, 2008). To do this they may have used appendages of the left side in detection of prey items, as well as danger (i.e. in general scanning).

Lateralized specialization of appendage sensillae is present in insects (discussed in Chapter 1). In the honeybee, there is right antenna advantage in learning to associate an odour with a food reward (Letzkus *et al.*, 2006) and in recall of short-term memory of that odour association (Rogers and Vallortigara, 2008). Electroantennography suggests that one contribution to this asymmetry is differing responsiveness of sensillae on left and right antennae (Anfora *et al.*, 2010). This may not be accompanied by marked changes in glomerular volume in the antennal lobes of the brain as a result of odour training (Rigosi *et al.*, 2011).

Recently, dominance of antennal input on one side in response to food smells has been reported for two other insects belonging to widely separated groups. In cockroach (*Periplaneta americana*) responding to food smells, turning towards the right is very strong when only the right antenna is functional, whereas there is much less clear bias to the left when it is the left antenna that is functional (Cooper *et al.*, 2011). In *Drosophila*, comparison of inputs to the two antennae is necessary for tracking odour plumes in the air (Duistermars *et al.*, 2009). However, when only one antenna is functional, the left still causes turning, which is to the left, suggesting that the source is then judged to be on the left; in contrast, there is little bias when only the right antenna is in use. *Drosophila* larvae use right and left groups of olfactory sensory neurons to locate food (Louis *et al.*, 2008). The larvae probably compare inputs obtained during sequential sampling in head sweeps. Genetic manipulation of development shows that retention of function on the right gives better orientation than retention on the left (i.e. the reverse of the adult lateralization).

Ganglionic specialization may also be associated with, and even depend on structural changes in, one of a pair of appendages, resulting from differences in use. In the lobster, direct experimental exercise of one of the pair of main claws causes it to develop 'crusher' properties, leaving the other to become the 'cutter' (Govind, 1992). Changes in coordination follow. It is thus likely that under natural conditions changes in use may direct the process of differentiation. Within crabs, some species show consistent direction of claw asymmetry (large or small) across a group of related species, whereas other species show no consistency even within a population (Duguid, 2010). Hence, there is, apparently, ready evolution of appendage asymmetry.

Lateralization of neural mechanisms at least sometimes depends on appendage ganglia. Independent learning in arthropods by appendage ganglia is fully established for the fruitfly, *Drosophila*, in which avoidance of shock by sustained retraction or extension of a leg can be learned in the absence of a head, presumably via the thoracic ganglia (Booker and Quinn, 1981; Tully and Quinn, 1985). Structures analysing other inputs (e.g. from other senses) clearly need to be informed about such local learning. In the honeybee, perceptual input from the antennae reaches directly only the ipsilateral antennal lobe in the brain, whereas input from the proboscis is routed bilaterally and mediates effects of reinforcement (Sandoz *et al.*, 2002). Here then, independent analysis of sensory inputs from right and left sides is apparently separately related to information about food cues.

The Nematoda resemble the Arthropoda in possession of a cuticle and in loss of cilia (Lartillot and Philippe, 2008). In *Caenorhabditis*, pairs of homologous olfactory neurons have been described (Sagasti, 2007; Ortiz *et al.*, 2009), which sense different sets of odorants through left and right members of the pair. However, this may be a special kind of right–left asymmetry in that the very small number of neurons in the central nervous system of *Caenorhabditis* may require left–right specialization to allow an adequate range of stimuli to be distinguished.

### 3.8.1 Memory formation

Lateralization appears to have a special role in memory formation and recall.

Learning by any complex nervous system requires extensive processing of information. Different sense organs will have acquired different information, which may be held by specific parts of the central nervous system quite separate from other structures that may already hold or will acquire relevant information. It may be advantageous to hold information relevant to a specific effector structure (e.g. in Arthropods an appendage) locally in the short term, when memory relevant to local conditions is likely to be more useful.

The existence of successive stages in the elaboration and recall of new memories has usually been considered only as reflecting the need to transform initial memory held in temporary neuronal states into permanent records, involving protein synthesis. However, the transition to long-term memory may also allow interaction between different relevant short-term and long-term memories to occur, without interference from sensory input due to ongoing behaviour. As a consequence of such interaction, linkages between memory traces or the deletion of ones inconsistent with established information might be possible.

In *Drosophila*, different lobes of the mushroom bodies (to some extent equivalent to vertebrate forebrain) are specially associated with short-term and long-term memory. The test used to show this was very similar to ones used in vertebrates, namely the acquisition of avoidance by choice in a T-maze of a specific odour, following association of the odour with electric shock. In other words, the *Drosophila* could choose not to enter the arm of a T-maze distinguished by a specific odour. Within the mushroom bodies, short-term, medium-term and long-term records are supported by changes that affect the same category of neuronal axons (Yu *et al.*, 2006). As in vertebrates, long-term memory requires protein synthesis. The transition from short- to long-term records of conditioning depends on an asymmetric body normally only present on the right side of the brain close to the structure connecting the two sides of the brain. In rare cases there is also a counterpart on the left, and in such flies only short-term memory is formed (Pascual *et al.*, 2004; see also Chapter 1). This suggests that information held in right and left side structures is normally in some way interrelated late in memory formation, with right and left sides taking different roles, and that this requires one side to take the lead.

In honeybees, there is a single large median neuron in the brain that mediates effects of reinforcement (sucrose), whether this is detected via antennae or proboscis (Sandoz *et al.*, 2002). When the conditioning stimulus is presented to one antenna, retrieval via the other develops progressively over intervals of 10 minutes to 3 hours. Involvement of the other antennal lobe, which did not initially detect the odour, may involve transfer of information or the establishment of linkages.

Following training with both antennae in use, bees show better recall using the right antenna for up to 2 hours after training, but by 6 hours and 24 hours the advantage shifts to the left antenna (Rogers and Vallortigara, 2008). The shift of recall from right to left shown by the honeybee presumably relates to shifts from foraging dominated by recent experience to reliance on long-term memories (Chapter 1). At the same time it is clear that long-term changes in olfactory structures (patterns of activation of olfactory glomeruli in antennal lobes), which differ between right and left sides, persist after such training (Hourcade et al., 2009). The shifts in recall between short-term and long-term memory seen in foraging may, therefore, reflect changes in access by higher structures to information held by the antennal lobes.

There is considerable evidence, in vertebrates, for memory formation in parallel on the right and left sides and for interaction between the left and right memory traces during memory formation (Chapter 5). It is likely then that the advantages of shifts from recently acquired information held to some extent separately on the right and left to more comprehensive and integrated long-term records are so great that mechanisms controlling such shifts have evolved (most probably separately) in both arthropods and vertebrates.

## 3.9 Conclusion

Bodily asymmetry was dominant in the early evolution of chordates (see also Andrew, 2002b) and became expressed as lateralization of central nervous system and behaviour throughout vertebrate species. Given its commonality, even in invertebrates, similar advantages of lateralization probably operated to some extent in chordates and other groups. In vertebrates the left hemisphere became specialized for the control of well-established patterns of behaviour under familiar circumstances, whereas the right hemisphere became specialized for detecting and responding to unexpected stimuli in the environment. These asymmetries eventually led to the evolution of speech and language control by the left hemisphere, probably by its specialization for routine communication, both vocal and non-vocal, and control of planned sequences of responses. The long-standing disputes about the origin of human language appear thus to be solved by an evolutionary sequence in which initial dominance of gestural communication controlled by the left hemisphere was followed by the addition of vocalizations.

# Development

## Summary

Experience can enhance, suppress or change in other ways the development of lateralization. Exactly which of these occurs depends on the species, the nature of the experience and the stage of life at which it takes place. Lateralization of individuals and groups can be modulated by experience and by steroid hormones. The latter may be important in the development of sex differences in lateralization. Research in this area is in its infancy compared with our knowledge of species differences in lateralization, but we are able to give some potent examples to illustrate the importance of experience and hormone levels at particular stages of development.

## 4.1 Introduction

The brain is not as hard-wired as once thought. It changes its connections in response to experience, especially in early life but also in adulthood. Some regions of the brain even change size in response to specific kinds of experience. The hippocampus is such a region. In humans, we know that the hippocampus has a special role in spatial memory. A study of London taxi drivers has shown that they have a larger than average posterior region of the right hippocampus and a smaller than average anterior region of the hippocampus (Maguire *et al.*, 2000). In animals too, the size of the hippocampus is related to spatial ability. Species that cache food and retrieve it at a later time have a larger hippocampus than do closely related non-caching species. This is known to be the case in squirrels (Johnson *et al.*, 2010), kangaroo rats (Jacobs and Spencer, 1994) and several species of birds, including marsh tits and Clarke's nutcrackers (Shettleworth, 2003).

It makes sense that the caching species have larger hippocampi because they have to make use of spatial memory to retrieve the food items that they have hidden, but having a larger hippocampus is not entirely due to having genes that programme for this. As we know from studies of marsh tits, if the birds do not have the opportunity of caching food, they do not have a large hippocampus (Clayton and Krebs, 1994a). Their brain changes its structure and connectivity as a result of the experience of caching. To put is another way, in order to have a large hippocampus the birds must have not only the genes of a caching species but also the experience of caching food.

Using spatial ability is important for the development of lateralization in certain regions of the brain in homing pigeons, as shown by Mehlhorn *et al.* (2010). These researchers compared homing pigeons that had or had not been given the opportunity to navigate by flying around their loft and participating in races from the time of fledging (at about 28 days post-hatching) until adult maturity. The pigeons that had been able to navigate had a number of asymmetries not present in those that had been unable to do so: the experience of navigation led to a larger left hippocampus, right nidopallium and right optic tectum. Hence, experience some time after hatching is important for the development of brain asymmetry.

Knowing this raises the question of whether other forms of brain lateralization can be altered by experience. The answer is yes. We know that at least some forms of lateralization are malleable depending on particular types of experience at particular stages during development. Different kinds of lateralization can be enhanced, suppressed and even generated by experience. Although genes lay the foundations for building a lateralized brain, experience can modify it and this is important because it allows individuals and species to adapt to changing environments. In some social contexts, for example, it may be advantageous to be lateralized, whereas in others not being lateralized may bestow advantages. This chapter presents evidence for the influence of experience on the development of lateralization. It also shows that experience can affect which hemisphere tends to be dominant in controlling behaviour. As a result, some animals behave consistently as if their left hemisphere is in charge of their behaviour, whereas others behave as if the right hemisphere is in charge. Stress, for example, seems to contribute to dominance of the right hemisphere and expression of this hemisphere's characteristic pattern of behaviour (Rogers, 2010a).

Especially in early life, experience can influence the development of lateralization. Absence of certain sorts of stimulation during specific periods of development can lead to failure of lateralization to develop. Two lines of evidence have demonstrated this: one has involved exposure of the developing embryo to light and the other has involved the experience of 'handling' during the neonatal period. Handling is a standard procedure used to stimulate rat pups and it involves separating them from their mother for a few minutes each day, usually over the first 21 days of life. Let us consider how these two very different forms of stimulation affect the development of lateralization.

## 4.2 Light and development of lateralization

During a particular stage of development, known as a sensitive period, exposure to light permits the development of neural pathways and structures in the brain so that processing of visual inputs is lateralized. Consequently, the animal responds differently to stimuli that it sees on its left and right sides. This has been shown in the zebrafish and in two avian species (domestic chicks and pigeons). In the zebrafish, light stimulation acts on an asymmetrical structure, the parapineal and its connected habenular nucleus, to cause the development of visual laterality. In birds, light-stimulated asymmetry of the visual system results from positioning of the embryo in the egg that allows only the right eye to be stimulated by light.

### 4.2.1 Light and lateralization in fish

Some of the lateralized functions of zebrafish develop only if the embryo is exposed to light. Fry hatched from eggs that have been exposed to the normal light/dark cycle for the first six days following fertilization use their left eye to view conspecifics and stimuli that might provide them with a refuge from predators (Andrew et al., 2009b). This particular lateralization is not seen in fry hatched from eggs that have been incubated in the dark (Andrew et al., 2009b). Light-exposed and dark-incubated fry also differ in their response to a model predator shown to them on their left or right side as they emerge through a small opening into a new enclosure. Fry that have been exposed to the light/ dark cycle during early development avoid the predator when it is presented on their left side but approach it when it is on their right side (Budaev and Andrew, 2009a). Fry that have developed in darkness show no such asymmetry: they avoid the predator regardless of the side on which it is shown. A comparable result was obtained by presenting a pattern of closely spaced vertical lines, which resembled vegetation and might be seen as a potential place in which the fish could hide. Fry that had developed in the light/dark cycle approached the stimulus when it was presented on their left side and not when it was presented on their right side, but fry that had developed in darkness showed no asymmetry (Andrew et al., 2009b). Zebrafish exposed to light during development also have higher scores of boldness than those that have developed in the dark (Budaev and Andrew, 2009a).

Although light exposure during early development strongly affects lateralization in zebrafish, some aspects of lateralization are still manifested in zebrafish reared in darkness. Heightened avoidance of a conspicuous novel object seen with the left eye is still present (Andrew et al., 2009b). Also, the pattern of eye use during learning about a mirror reflection for the first time shows asymmetry in both dark- and light-reared zebrafish, but of a different character

(Andrew et al., 2009b). At times of rapid change in memory formation, light-reared fry shift very rapidly from left to right eye viewing of their reflection seen for the first time, whereas dark-reared fry show left eye use on the first occasion and then right use on the second occasion (Andrew et al., 2009a). Thus the right side of the brain (left eye) appears to have initial priority in examining a complex and valent stimulus and then the left hemisphere takes over control. Some aspect of this control is reduced in dark-reared zebrafish.

Exposure of the developing fry to shorter periods of light and dark has revealed two sensitive periods during which light stimulation affects the development of lateralization. The first day after fertilization marks one sensitive period during which absence of light exposure, somewhat surprisingly, causes reversal of the usual asymmetry of response to a model predator: these fish now respond more strongly to the predator when it is on their right side and they persist in avoiding it (Budaev and Andrew, 2009b).

The reversal of facilitated avoidance of a model predator from the left to the right eye, by the absence of light on day 1, shows that light has a crucial effect at this time (Budaev and Andrew, 2009a, 2009b). Changes in habenular connectivity are probably involved. Each side of the diencephalon has a habenular nucleus and each habenular nucleus has more than one lobe, each differing in size: the lateral lobe of the habenula is larger on the left side than is its counterpart on the right side, and the medial lobe is larger on the right side than on the left (Aizawa et al., 2005; Halpern et al., 2005). The left lateral normally opposes intense avoidance (Agetsuma et al., 2010, and discussed further in Chapter 5). Concha et al. (2009) showed that stem cells for habenular neurons are of two types. In one, 'left typical', the marker lov appears earlier and chiefly on the left; in the other, 'right typical', the marker is ron and here the appearance is later and chiefly on the right. The left-typical neurons are born as early as 24 hours after development begins, whereas right-typical neurons do not begin to appear until 48 hours (Aizawa et al., 2007). Since an effect as early as 24 hours cannot involve photoreceptors, which are not yet present, activation of genes by light is the probable route for this effect. It is, in fact, known that light exposure at this stage activates the expression of the gene nrp1a that guides the growth of axons from the left lateral habenula to another brain region known as the interpeduncular nucleus, IPN (Kuan et al., 2007b). It is possible that failure of activation by light could result in the left lateral habenula driving the normal outflow route of the right medial habenula. The intense avoidance shown after absence of light on day 1 suggests that here the left habenula is sustaining the response of avoidance, much as it normally sustains responses such as pursuit of prey (Chapter 3).

The second sensitive period is on the third day after fertilization. Fry that receive no light stimulation on this day fail to develop lateralization of responses to the model predator (Budaev and Andrew, 2009b). This suggests some general failure of habenular outflow, perhaps associated at this time with absence of light-induced drive of the parapineal.

It is clear that stimulation by light has different effects during the two sensitive periods, day 1 and day 3. In order to develop the usual pattern of lateralization, the fish must be exposed to light on both these days. On the third day after fertilization it seems that light affects the development of asymmetry in the zebrafish by stimulating a light-sensitive organ in the diencephalon of the brain, the parapineal, and the habenular nuclei, to which it is connected. The connections of the parapineal to the habenular nuclei are asymmetrical in some species (Engbretson et al., 1981; Halpern et al., 2003; Guglielmotti and Cristino, 2006); the light-sensitive cells connect mainly to the left habenular nucleus. In the zebrafish these particular connections appear to be symmetrical (Yánez et al., 2009) but connections making up the next stage, from the habenular nuclei to the interpeduncular nuclei of the midbrain, are asymmetrical (Kuan et al., 2007a). This means that the effects of stimulation by light, which readily reaches the embryo because the eggs of zebrafish are transparent, are asymmetrical. Because the left and right habenular nuclei project to different regions of their respective interpeduncular nuclei (Kuan et al., 2007a), light stimulation could trigger widespread, asymmetrical changes in neural connections in the brain (Andrew et al., 2009b).

Since the habenular nuclei are asymmetrical in many other species (Braitenberg and Kemali, 1970), it is possible that lateralization in these species too relies on light stimulation during embryonic development. As in zebrafish, in frogs, newts, lizards and eels the left habenular nucleus is more lobate (summarized in Bradshaw and Rogers, 1993) and the left habenular nucleus responds to light (Guglielmotto and Cristino, 2006), providing an asymmetrical substrate on which light may act to enhance lateralization or generate lateralization in other brain regions (Andrew et al., 2009b). This has not yet been investigated in detail.

Such a role of light in the development of lateralization could interface with predation on the species. Depending on levels of predation, fish might lay their eggs in places where they receive more or less exposure to light during the sensitive periods and this would influence the degree, and even the direction, of lateralization developed by the offspring. In fact, poeciliid fish captured from ponds in which predation levels are high are more lateralized than the same species captured from ponds, in the same stream, with lower levels of predation (Brown et al., 2004). But differences in experience are not the only cause of the difference in lateralization between fish from high and low predation ponds because fish caught from these two locations and allowed to breed in controlled laboratory conditions produce fry that are similar to their parents in strength of lateralization, although not necessarily in the direction of lateralization (Brown et al., 2007). Clearly some genetic selection has taken place even though light exposure, or some other stimulation, might have been influential in the initial divergence of the two populations and might still have some influence on the development of lateralization.

### 4.2.2 Light and lateralization in birds

Light stimulation during the development of the avian embryo also affects the development of lateralization but the mechanism differs from that demonstrated in the zebrafish. In birds, it is asymmetrical stimulation of the eyes (i.e. stimulation of one eye and not the other) by light during the final stages of development before hatching that leads to the development of lateralization of certain patterns of behaviour dependent on vision. During the final few days of incubation, the embryo fills the available space inside the egg and turns its head so that its beak points to the right, and the left eye rests next to the embryo's body (Figure 4.1), probably as a result of unilateral expression of genes such as *Shh*, *lefty*, *nodal* and *Pitx2*, known to determine the side on which the heart, gonads and some other organs develop (Levin *et al.*, 1995; Guioli and Lovell-Badge, 2007). By adopting this position the embryo occludes its left eye with its own body. The right eye remains facing towards the shell, through which some light can penetrate. Since the embryo's head is almost always to the left side (e.g. in 99% of domestic chicks; Kovach, 1968), stimulation of the right eye by light has a strong population bias.

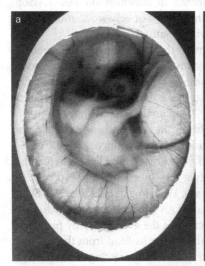

Figure 4.1 Chick embryos at two stages of development. **a**, Embryo on day 8 of incubation. Note that it is lying with its left side against the yolk sac and with the right eye next to the shell and membranes. The visual system is not functional at this stage. **b**, One day before hatching the visual system has become functional and the embryo's head is turned to the left side. The left eye is occluded by the embryo's body and the right eye is positioned next to the membranes and eggshell. Hence, only the right eye can be stimulated by light. Note that the membranes of the air sac and the eggshell have been removed in **b**. Published previously in Rogers (1995).

In domestic chicks, the retina and visual projections to the forebrain (pallium) have become functional by the time the embryo turns its head to the left side. Therefore, in response to stimulation by light, the visual regions of the brain receiving input from the right eye develop in advance of those receiving input from the occluded, left eye (Rogers and Sink, 1988; Rogers and Deng, 1999). As a consequence, the right eye establishes more connections from the thalamus to the forebrain than does the left eye.

Birds have two visual pathways to the forebrain: the thalamofugal and tectofugal pathways (Figure 4.2). In the chick (*Gallus gallus*), the one developing lateralization in response to light exposure of the embryo is the thalamofugal pathway. Structural asymmetry in the thalamofugal visual pathway of the chick persists for the first 3 weeks of life after hatching. This seems to allow ample time for this asymmetry of visual processing to lead to the development of many lateralities of behaviour, some of which persist into adulthood (e.g. attack responses; Rogers, 1991).

Right eye superiority in inhibiting pecks at pebbles while searching for grain and left eye superiority for predator detection, attack and copulation all depend on light stimulation of the embryo (Rogers, 1982, 1990, 2000). We know this because chicks hatched from eggs incubated in darkness do not develop asymmetry in the performance of any of these types of behaviour and they do not have asymmetry of the thalamofugal visual pathway.

The influence of light on the development of these several types of lateralization is quite specific: there are no other known effects of light exposure versus dark incubation on performance. For example, a detailed comparison of the kinematic aspects of walking in light-exposed versus dark-incubated chicks failed to find any differences in performance (Sindhurakar and Bradley, 2010). Although it is known that exposure to light can induce hatching to take place one or two days earlier than it does in dark-incubated embryos (Siegel *et al.*, 1969), this effect is seen only after prolonged exposure to bright light, whereas the effect of light on lateralization requires only a very short period of exposure to relatively low levels of light (Zappia and Rogers, 1983; Rogers, 1990, 1999a). Such a small amount of light stimulation has no obvious effect on time of hatching.

In fact, it is possible to reverse the direction of the asymmetry by placing a patch over the embryo's right eye after withdrawing its head from the egg just one or two days before hatching will take place. Withdrawing the embryo's head from the egg at this stage of development does not affect the chick's survival because a chick embryo of this age can breathe air, which it normally does from the air sac end of the egg (Rogers, 1995). By putting a light inside the incubator the manipulated embryo receives stimulation via its left eye and not via the patched, right eye. The result is reversed asymmetry of the visual pathways and of visual behaviour (summarized by Rogers, 2008). A control was necessary to show that the reversal had not been caused by non-specific effects of the procedure. That control involved the same withdrawal of the

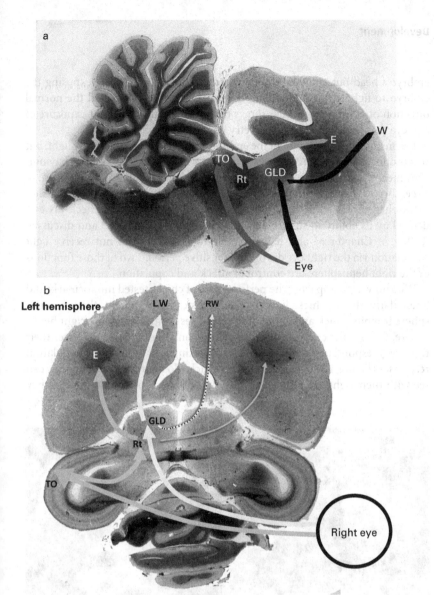

Figure 4.2 The visual pathways of birds. Panel **a** is a sagittal section of the chick brain showing the tectofugal (grey lines) and thalamofugal pathways (black lines). The chick's forebrain is to the right side and the hindbrain and cerebellum to the left side. Note that this section does not illustrate the asymmetry. Panel **b** is a horizontal section of the chick brain showing only the thalamofugal pathway, which receives input from the right eye. The mirror image projections from the left eye are not represented. Note the projections from GLD to the Wulst in the opposite (right) hemisphere, represented by the white line with black dots. The asymmetry is located in these projections: there are more projections from left GLD to right Wulst than from right GLD to left Wulst. Abbreviations: TO, tectum opticum or optic tectum; Rt, nucleus rotundus; E, entopallium; GLD, nucleus genticulatis lateralis pars dorsalis; LW, left Wulst or hyperpallium; RW, right Wulst or hyperpallium.

embryo's head but applying the eye-patch to the left eye before exposing the embryo to light. After undergoing this treatment, the chicks had the normal direction of asymmetry, which shows that it is definitely the light exposure of one eye that leads to visual asymmetry after hatching.

During normal development the embryo's right eye is stimulated by light, but this monocular experience alters the way in which both eyes guide behaviour. This means that the function of both hemispheres is organized by exposure of the embryo's right eye to light. It is most likely that this happens because the left hemisphere holds some functions of the right hemisphere in check (as evidenced by its ability to sustain an initiated response; Chapter 3 and discussed further in Chapter 5) and it is less able to this if it does not receive light stimulation via the right eye during the sensitive period. Two of those functions of the right hemisphere are control of attack and copulation.

We know, by comparing the performance of chicks tested monocularly and binocularly, that in chicks hatched from eggs exposed to light, the left hemisphere inhibits attack and copulation responses controlled by the right hemisphere. Chicks that can see with their left eye only (i.e. with a patch over their right eye) respond to a moving hand by attacking it or performing copulation responses (Figure 4.3), whereas the same chicks do not respond when they can see with their right eye only. They also do not attack or copulate when they

Figure 4.3 Attack and copulation. Chicks responding to a moving hand by attacking (a) or copulating (b). Both chicks have been treated with testosterone to elevate the likelihood of eliciting this behaviour. By scoring the chick's responses, ranging from 0 for turning away to 10 for full attack or juvenile copulation, the level of responsiveness can be measured. Panel a shows the chick lifting one foot in preparation for an attack leap at the moving hand. Panel b shows a chick that had mounted the hand, adopted a crouched posture and is performing pelvic thrusts as it is treading. Note that even untreated chicks perform elevated levels of attack and copulation provided that (1) they have been exposed to light before hatching and (2) they are tested with the right eye covered. They rarely attack or copulate if the left eye is covered or if they are tested without any eye-patches.

can see with both eyes. This shows that the left hemisphere (right eye) is dominant in young chicks and suppresses responses of the right hemisphere. Exposure to light before hatching is essential for development of this left hemispheric dominance: chicks hatched from eggs that have not been exposed to light show the same level of attack and copulation when they can see with either the left or right eye. Also, exposure of the embryo's left, and not the right, eye to light reverses the asymmetry of attack and copulation responses (Rogers, 1990).

Light exposure of the chick embryo also promotes the ability to combine position with colour cues. Chiandetti and Vallortigara (2009) trained chicks to find a food bowl in the corner of a cage with a blue wall on the chick's right side and a white wall on its left side. Then the chicks were tested with a choice between a corner with this arrangement of the walls and a corner with the blue and white walls reversed. Chicks exposed to light before hatching approached only the bowl with blue on the right and white on the left, but chicks that had not received light exposure before hatching could not distinguish between the two types of corner. Although they could still discriminate colour, they failed to combine position with colour.

Response to a potential predator is another function of the right hemisphere that is held in check by the left hemisphere, provided that the left hemisphere (right eye) has been exposed to light before hatching. As explained in Chapter 2, when chicks hatched from eggs that have been exposed to light are searching for food and then a model predator flies overhead, they stop pecking to look at the predator but soon return to feeding. Their left hemisphere assumes control, allowing them to inhibit responding to the predator and go back to searching for food. Chicks hatched from eggs incubated in darkness are less likely to detect the predator in the first place but, once they have seen it, they are more distracted from feeding and become quite disturbed (Rogers et al., 2004; Wichman et al., 2009). They seem to be locked into using the right hemisphere, shown by the fact that they look at the predator with the left eye (Rogers, 2011a, 2011b). Absence of light exposure of the right eye appears to have reduced the left hemisphere's ability to suppress the fear expressed by the right hemisphere. This result is consistent with the finding that exposure of the embryo's right eye to light allows the non-stimulated (right) hemisphere to respond effectively to novel stimuli (Chiandetti et al., 2005).

Very brief exposure of the chick embryo to light is sufficient to bring about lateralized visual functions. As little as two hours can be sufficient as long as it takes place between day 19 of incubation and hatching on day 21 of incubation (Rogers, 1990). This sensitive period marks the stage of development when the visual pathways to the forebrain are becoming functional and, as said above, it is the one projecting from the thalamus to the forebrain that is influenced by the asymmetrical stimulation of the left and right eyes. Such experience-dependent asymmetry may apply generally to precocial birds, meaning birds that hatch at a relatively advanced stage of development and can walk and forage almost immediately.

Light also influences the development of lateralization in an altricial species, the pigeon. Altricial species are hatched in an immature stage and require feeding and other care while they remain in the nest. The visual lateralization of the pigeon depends on exposure to light before hatching but it develops in the tectofugal, not the thalamofugal, visual system. The visual projections from the optic tectum on each side of the brain to the rotundal nuclei (Figure 4.4), in the midbrain and on each side of the brain, are asymmetrical in the pigeon. The light-dependent asymmetry develops in the cell sizes of particular neurons of the optic tecta (Güntürkün, 1993), in the neurons projecting from the optic tecta to rotundal nuclei (Güntürkün, 2002) and in neurons in other areas connected to the tecta (Freund et al., 2008). The rotundal nucleus on the left

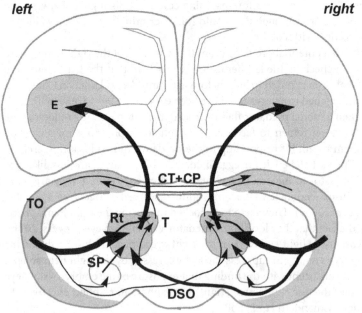

Figure 4.4 Visual pathways of the pigeon. The asymmetrical tectofugal system is represented. Shown in this brain section are the forebrain hemispheres (at the top) and the diencephalon (midbrain). Abbreviations: TO, optic tectum, which receives direct inputs from the opposite eye; Rt, rotundal nucleus; E, entopallium; T, nucleus triangularis; SP, nucleus subpretectalis; CT+CP, tectal and posterior commissures; DSO, dorsal supraoptic decussation. The main pathway is from eye to TO, to Rt, to E. Note the asymmetry in the TO to Rt projections that cross the midline in DSO: there are fewer from left TO to right Rt than *vice versa*. Figure courtesy of Professor Onur Güntürkün.

side of the midbrain, which receives input from the optic tecta and then projects to the forebrain, receives almost equal numbers of neural projections from both the right and left optic tecta. The right rotundal nucleus receives input mainly from the right optic tectum. This means that the left hemisphere receives visual inputs from both eyes, unlike the right hemisphere, which has input predominantly from the left eye. Furthermore, the left optic tectum is better able to inhibit its partner on the other side of the brain than *vice versa*. Hence, once inputs from the right eye become the focus of attention, such control can be sustained more readily than would be the case for inputs from the left eye. In other words, pigeons tend to be right eye dominant.

Such dominance of the right eye system does not occur in chickens. Since the tectofugal system is specialized to detect moving targets, chickens should be more responsive to, say, insect prey detected on their left or right side than would seed-eating pigeons with their right eye dominance.

Once light has caused these changes in the pigeon, the lateralization persists into adulthood, and it involves connections both to and from the forebrain (Valencia-Alfonso *et al.*, 2009). Adult pigeons peck at food grains and avoid pecking at grit (pebbles) when they use their right eye but not when they use their left eye. Pigeons hatched from eggs incubated in the dark do not have these asymmetries (Güntürkün, 2002).

The fact that light exposure of the embryo triggers development of asymmetry in the tectofugal visual system of the pigeon but not in the tectofugal system of the chick (Rogers and Deng, 1999) could be explained by one species being altricial and the other precocial (Deng and Rogers, 2002a). The tectofugal pathway becomes active before the thalamofugal, as found by uptake of the radioactively labelled sugar 2-deoxyglucose into neurons that are active. In chicks the tectofugal system becomes active well before the embryo moves into the turned position occluding the left eye (Rogers and Bell, 1989). Only the thalamofugal system is becoming active at the critical stage just before hatching when the chick embryo turns its head to the left side and the right eye is stimulated by light. By contrast, pigeons and other altricial species adopt the turned position at a stage of development when the tectofugal system is becoming active. As a consequence, the tectofugal system is stimulated asymmetrically by light and develops asymmetry.

Asymmetry of the thalamofugal system in the pigeon is either absent or much weaker than in the chicken. Some evidence shows that the region of the forebrain receiving inputs from the thalamofugal system is asymmetrical in volume (Mehlhorn *et al.*, 2010), which may suggest some asymmetry in the thalamofugal projections, but recent investigation of the thalamofugal system of the pigeon has revealed no significant asymmetry in newly hatched birds or in adults (O. Güntürkün, personal communication, February 2012). Moreover, a period of monocularity just after hatching failed to generate any asymmetry in the thalamofugal pathway. Therefore, it is certain that pigeons and chickens have asymmetry in different visual pathways. Consequently, some differences in their lateralized behaviour are expected (discussed in Chapter 5).

Chickens and pigeons also differ in that asymmetry in the thalamofugal pathway of the chickens is transient, disappearing by 3 weeks after hatching (Rogers and Deng, 1999), and asymmetry in the tectofugal pathway of the pigeon is permanent (Güntürkün, 2002). However, since adult chickens display asymmetry of attack responses (Rogers, 1991), it seems that the transient asymmetry of the thalamufugal pathway triggers more permanent asymmetries in other parts of the chicken brain.

Other differences in the role of light exposure in establishing visual lateralization are expected between avian species. For example, birds that nest in burrows, mounds or dark tree hollows are not likely to be exposed to light at a stage when monocular exposure is matched by the turning of the embryo's body. In fact, embryos of megapodes do not adopt the turned posture before hatching, and two megapode species, the Australian brush turkey (*Alectura lathami*) (Figure 4.5) and mallee fowl (*Leipoa ocellata*) (Baltin, 1969), lay and incubate their eggs in mounds of earth and rotting leaves. Their chicks hatch underground and dig their way to the surface before they are exposed to light, at

Figure 4.5 An Australian brush turkey, *Alectura lathami*. This species lays its eggs in a mound of vegetable matter, which acts as a natural incubator as decomposition takes place. The eggs are not exposed to light before hatching. Photograph by Professor L. Rogers.

which stage both eyes are open and there is no possibility of the monocular stimulation required to develop asymmetry of the visual pathways. Consequently, they may have no lateralization of the visual pathways or of those kinds of visual behaviour discussed above, although they may well be lateralized for other non-visual behaviour. Alternatively, any asymmetries in their visual systems, if they exist, may depend on left–right differences in gene activation. This has not yet been examined.

These examples show that light has a special role in the development of visual lateralization in birds and in fish, but there are species differences in the mechanisms involved and in whether or not the eggs are exposed to light. Perhaps more important is the fact that intra-species differences may result from differing conditions of light exposure of the eggs. Food availability could influence the time during which the incubating bird leaves its nest and, in turn, the amount of light exposure of the eggs. Level of predation may be an influential factor in this context since parents leave the nest more or less often depending on predation. Offspring of parents that have stayed on the nest continuously over the final days of incubation might have enhanced responsiveness to predators owing to lack of exposure to light during the sensitive period. This may influence their survival after hatching, depending on the kind of predation and other factors in their environment. The same should be the case in species that incubate their eggs in dark places (e.g. bee-eaters and the rufous hornero of South America; Figure 4.6) and birds with opaque eggshells (e.g. emus; Figure 4.7).

Chance variation in exposure to light might also produce variation between individuals in characteristics such as boldness. Such variation could be advantageous when there are large numbers of offspring (e.g. as in some fish).

## 4.3 Handling and development of lateralization

Light exposure affects development of laterality in species that develop in eggs. In mammals any role of light in the development of lateralization is less likely, although it has been said that some light could reach the foetus developing in the uterus and it has been shown that the human foetus can respond to light. Kiuchi et al. (2000) shone a flashlight on the abdomens of pregnant mothers and found that some of the foetuses, at 36–40 weeks' gestation, changed their heart rate and movement of their eyes and body in response to the light stimulation. These researchers also applied vibroacoustic stimulation to the abdomen and found that it was a more effective stimulus but, nevertheless, light did have an effect provided that the foetus was awake.

Lateralized stimulation of one eye of the foetus more than the other is not inconceivable because, as we know in humans at least, just before birth, the

Figure 4.6  Nest of the Brazilian rufous hornero (*Furnarius rufus*) or Ovenbird. Eggs incubated in this nest, constructed of mud, would receive very little, if any, exposure to light. Note the internal wall shielding the main part of the nest from the opening. Photograph taken by Professor L. Rogers in the Pantanal, Brazil.

foetus' head is locked into the mother's pelvis and, in two-thirds of cases, it is turned towards the mother's right side (see Figure 8.2 in Hellige, 1993a). Infants that have adopted this position prefer to turn their head to the right side after birth and they show right-hand preferences in infancy (Michel and Goodwin, 1979; Michel, 1981). When the foetus is in this turned position, it is not entirely impossible that one eye might be stimulated by light more than the other and so have some influence on the development of hemispheric specialization.

Nevertheless, such an effect of light seems rather unlikely. Alternatively, as proposed by Previc (1991), this preferred direction of turning of the foetus, together with the fact that humans walk forwards most of the time, would lead to lateralized stimulation of the vestibular reflexes of the foetus and that could influence the development of lateralization. According to this theory, the left otolith, the organ of the inner ear responsible for balance, receives more stimulation than the right and this leads to the development of the preference of humans (cf. other primates: MacNeilage *et al.*, 1987) to use the left side of the

Figure 4.7 Eggs of three species. From top to bottom: emu, *Dromaius novaehollandiae*; domestic chicken, *Gallus gallus domesticus*; zebra finch, *Taeniopygia guttata*, showing variations in shell transparency/opacity. Emu eggs are almost opaque. Domestic chicken eggs vary from white to the brown colour seen here. Although brown eggshells transmit less light than do white eggshells, they allow a considerable amount of light to enter the egg. Eggs of zebra finches are translucent. See Rogers (2011b) for measurements of the amount of light entering the eggs through these different shells. Photograph by Professor L. Rogers.

body for postural control. It would also free the right limb(s) for voluntary motor behaviour, as used to manipulate objects, and it might even lead to the right hemisphere advantage in visuo-spatial processing (Previc, 1991). Although some circumstantial evidence supports this theory, it has not yet been tested empirically.

Experimental evidence has shown that one particular type of experience, handling, in neonatal life affects the development of lateralization in rats.

Handling unmasks the specialized functions of the right hemisphere (Denenberg, 1984). As mentioned at the beginning of this chapter, handling is the term used to describe removing each pup from its mother for a few minutes daily, usually over the first 21 days after birth. The term suggests that the main effect of this procedure might be tactile (touch). Not only are the pups handled by humans but also, after they have been returned to their mother, she licks them more than normal (Smotherman et al., 1977), thereby providing increased tactile stimulation. However, other kinds of stimulation could be involved. For example, the pups are distressed during the period of separation from their mother. Unfortunately, the means by which handling affects the development of lateralization has not been fully investigated, although Denenberg (2005) has suggested that the chief operative may be the stress suffered.

At least one of the effects of handling is to change the amount of communication between the hemispheres: pups that are handled develop a larger corpus callosum (the tract that connects the left and right hemispheres) than do pups that are not handled (Cowell and Deneberg, 2002). The corpus callosum is involved in competitive interactions between the hemispheres and in sharing information between the hemispheres. During development, the connections made by the neurons with axons in the corpus callosum are reorganized and, in general, some of the connections of the neurons are pruned (summarized in Bradshaw and Rogers, 1993). Various sorts of sensory stimulation affect this process. Handling may result in a larger corpus callosum either by preventing attrition of neurons or by stimulating growth of neurons, or by both processes.

Handling in neonatal life may well be important for the development of lateralization in many mammalian species. So far, the only other species investigated is the horse. Foals touched by humans on the right side in early life were found to be more wary of humans than were foals touched on the left side or not touched at all (de Boyer des Roches et al., 2011). In this case, the tactile stimulation is itself asymmetrical and so has something in common with the lateralized light stimulation of the avian embryo. It is possible that quality of experience for the horse differs according to the side touched.

Another effect of handling, shown in rats, is to change the asymmetry of the size of the hippocampus: the size of the right hippocampus is increased (Verstynen et al., 2001).

If stress is instrumental in the effect of handling on development of lateralization in the mammalian brain, one may ask by what means it has this effect. One potential answer is via the hormones released when stress is experienced. In fact, stress hormones are amongst a range of steroid hormones that are known to affect the development of some aspects of lateralization, as discussed below.

## 4.4 Hormones and lateralization

### 4.4.1 Stress hormones

Corticosterone is the main stress hormone in birds. Elevated levels of this particular steroid hormone during the sensitive period to light just before hatching prevent the development of asymmetry in the visual pathways (Rogers and Deng, 2005) and alter the chick's responses to predators (Freire et al., 2006). This was discovered by injecting a slow-release form of the hormone into eggs just before the sensitive period to light stimulation. By this means, corticosterone levels are abnormally high during the stage of development sensitive to light exposure. The experiments showed that light exposure of the embryo interacts with the hormonal state in determining the development of lateralization.

There are two ways in which corticosterone could become elevated naturally during this period. One is via the embryo experiencing stress, possibly as a result of becoming too cold, and releasing the hormone itself. Another way is for the egg yolk to contain hormone deposited there from the hen's blood stream as the egg was forming in her body. In fact, research on canaries indicates that hens deposit less corticosterone in each egg as it is laid in sequence in the clutch (Schwabl, 1999). It is conceivable, therefore, that canary chicks hatched from eggs laid later in the clutch may have stronger lateralization than do those laid earlier. However, species differ in whether corticosterone levels decrease or increase with order of laying: in wild European starlings, corticosterone levels increase across the laying sequence (Love et al., 2008).

These consistent trends of increasing or decreasing levels of corticosterone (depending on species) in accordance with order of laying in the clutch might affect brain development and might lead to consistent differences in behaviour after hatching. Competition between siblings to gain food from the parent might well be affected. In some species, order of laying and order of hatching are related, in which case it could be a distinct advantage for later-hatching offspring to be more competitive.

Even though order of laying is not always related to order of hatching, because the latter is often synchronized in the clutch (discussed in Rogers, 1995), eggs laid later may have fewer nutrients, and weaker young might hatch from them. In species in which sibling rivalry is intense, and especially those in which siblicide occurs, it could be an advantage for the weaker offspring to have more competitive behaviour. If we can extrapolate the evidence showing that domestic chicks with weaker lateralization (because they were hatched from eggs incubated in the dark) compete more strongly for access to food than those with stronger lateralization (hatched from eggs exposed to light) (Rogers and Workman, 1989), this might mean that smaller, weaker offspring would have more chance of survival if they are less lateralized.

## 4.4.2 Sex hormones

Two other steroid hormones have effects on the development of lateralization. They are the sex hormones oestrogen and testosterone. As in the case of corticosterone, in chicks, elevated levels of either of these hormones during the sensitive period to light prevent the development of lateralization of the visual pathways (Schwarz and Rogers, 1992; Rogers and Rajendra, 1993). Although the effects of varying doses of these hormones have not yet been investigated, it is reasonable to assume that differing levels of oestrogen and testosterone during the final stages of incubation may lead to sex differences in lateralization in chicks. In fact, the differing levels of these hormones may explain why the visual pathways of female chicks are less lateralized than those of male chicks (Adret and Rogers, 1989; Rajendra and Rogers, 1993). Also, as discussed for corticosterone, the level of androgens (testosterone and other steroids) varies with order of laying and, in turn, with order of hatching and level of aggression. This has been shown in black-headed gulls (*Larus ridibundus*); androgen levels increase with order of laying, which may facilitate growth and competitiveness of the later laid offspring (Groothuis and Schwabl, 2002).

There has been a long history of belief that sex hormones cause the known sex differences in lateralization in humans. For example, Norman Geschwind and Albert Galaburda (1987) proposed that testosterone delays the development of certain regions of the left hemisphere in men, who consequently rely more on the functions of the right hemisphere than do women. This, they suggested, may explain the higher incidence of left-handedness and language disorders in men than women. While interesting, this hypothesis has proved difficult, or impossible, to test in humans.

Some of the proposed ways in which the hormones act could be tested in non-human animals now that we know they too have sex differences in lateralization. A complication is that any effect of these hormones is not entirely due to them acting in their own right but rather in interaction with experience. In birds, as shown in chicks, the levels of the steroid hormones interact with the exposure to light. Indeed, this simple sentence covers a range of possible interactions, since corticosterone, oestrogen and testosterone can all reduce light-stimulated asymmetry, and each hormone may do so via a different mechanism. The precise balance of the hormonal mix in the embryo will determine the outcome on lateralization. Added to this, since some steroid hormones block the effects of others, it is not a simple matter of saying that, for example, males have higher levels of testosterone and this explains why they have asymmetry different from that of females. This applies to mammals, including humans, as well.

Sex differences in behaviour are seen in a wide number of species, including rats. In rats the cross-sectional area of the corpus callosum is larger in males than females (Berrebi et al., 1988) and this has been shown specifically in a sub-region known as the splenium (Nünez and Juraska, 1998). This difference may result from the interaction of sex hormones with stimulation in early life. For example, female rat pups that had been injected with testosterone four days after birth and were handled developed a larger corpus callosum than both untreated females and testosterone-treated females that were not handled (Fitch et al., 1990; Denenberg et al., 1991). Clearly, the effect of the hormone must interact with experience in early life. This interaction is shown in males too: handling of male pups leads to the development of a larger corpus callosum than in non-handled males, but handling has no effect on the size of the corpus callosum in castrated male pups (Fitch et al., 1991). Since the castrated pups would have had negligible amounts of testosterone circulating in their blood stream, it is clear that handling has an effect only if steroid hormones are circulating in the blood stream.

Added to this, it has been observed that handled rat pups open their left eye before their right, whereas non-handled ones have no preference (Smart et al., 1986). Since handling would allow a period of visual stimulation of the left but not the right eye, it might generate visual lateralization in much the same way as in the chick. This observation in rats has, so far, not been pursued further but it offers a potentially fruitful comparison of the species, and the same mode of development may be widespread amongst mammals, since an early observation noted that rabbits open the right eye first (Narang, 1977) and the optic fibres of that eye become myelinated in advance of those from the left eye (Narang and Wisneiwski, 1977). In female Mongolian gerbils the right eye tends to open first but in males it is the left eye (Clark et al., 1993; Ryan and Vandenbergh, 2002). Although it is not known how important asymmetry of eye opening may be, some of the effects of light on the development of lateralization shown in birds may still be present in mammals. As in birds also, asymmetrical stimulation may be instrumental in the development of sex differences in behaviour but, of course, steroid hormones may have other effects on the development of sex differences, especially given that receptors for them are found in many parts of the brain and they differ in distribution in males and females (e.g. in foetal rhesus monkeys; Sholl and Kim, 1990).

So far, we have discussed the postnatal effects of hormones and experience on lateralization in rats. Prenatal effects are also seen. If pregnant female rats are stressed, their male offspring are less lateralized and their female offspring more lateralized than offspring of non-stressed rats (Alonso et al., 1991). Note that postnatal handling, which is considered to be stressful, has the opposite effect on males (it enhances lateralization). The difference between the pre- and postnatal effects appears to be the age at which the stress occurs. Since stress causes elevation of cortisol levels in mammals, it could be this hormone that acts directly on the developing brain to alter lateralization. Cortisol could also interact with the sex hormones to mediate effects on brain lateralization.

Prenatal effects on development of lateralization are also suggested by some recent research. For example, elevated testosterone in the male, human foetus increases the number of connections in the isthmus region of the corpus callosum (Chura *et al.*, 2010), which may increase the degree of interaction between the hemispheres. This result is opposite to that proposed by Witelson and Nowakowski (1991); they suggested that increased testosterone levels in the foetus would decrease the number of corpus callosal connections and lessen the interaction between the hemispheres, hence causing sex differences in humans. The latter hypothesis is not supported by studies of the influence of steroid hormones on lateralization in non-human mammals (Pfannkuche *et al.*, 2009), or on humans (Chura *et al.*, 2010). However, the corpus callosum has many sub-regions and this needs careful consideration.

## 4.5 Shifts in hemispheric dominance and later influence of experience

As the above examples illustrate, in both the rat and the chick lateralization can be modulated by the interaction of hormones and experience. The outcome depends on the hormonal mix, the stage of development and the type of stimulation. In fact, changes in circulating levels of hormones might even have transient effects on hemispheric dominance. An example of this is the effect of testosterone on attack and copulation behaviour of chicks a week or two after hatching. Young chicks rarely show attack and copulation responses unless they have been treated by testosterone or another androgenic steroid. This is the case when they are tested binocularly but, as said above, even untreated chicks will attack and copulate if they can see only with their left eye. Testosterone-treated chicks also attack and copulate when they can see only with their left eye but not if they can see only with their right eye (Rogers *et al.*, 1985). The difference between control and testosterone-treated chicks is that testosterone-treated chicks also attack and copulate when they can see with both eyes, whereas untreated (control) chicks rarely do so (Table 4.1). This shows that the likely way in which the hormone acts to elevate attack and copulation in young chicks is to release the right hemisphere's control of these responses from inhibition by the left hemisphere. In other words, the hormone has caused a shift of hemispheric dominance from left to right.

Shifts in hemispheric dominance seem to be characteristic of the developmental process. In chicks we know that the left hemisphere assumes control of behaviour over most of the first week of post-hatching life with a peak of strong control on day 8. Then the right hemisphere takes over briefly on days 10 and 11 post-hatching (Rogers, 2010b; Figure 4.8). Workman and Andrew (1989) have shown that these shifts in dominance are accompanied by marked appearances and disappearances of behaviour associated with the expression of the

**Table 4.1** Attack and copulation in chicks

| Eye(s) not covered by patch during testing | Untreated (control) chicks | Testosterone-treated chicks |
|---|---|---|
| Left | +++ | ++++ |
| Right | – | + |
| Left and right | – | +++ |
| Hemisphere dominant when both eyes can see | Left | Right |

– indicates no response.
+ indicates level of attack or copulation.
See Rogers (2011b) for actual scores.

dominant hemisphere as the chick passes through the various stages of development. The left hemisphere is in control when the one- or two-day-old chick imprints. The left hemisphere is also firmly in control when, on day 8, the chick begins to actively search for food and ingest it. On day 10, the beginning of the period of right hemisphere control, chicks begin to go out of sight of the hen for the first time and commence to spar and perch. They also start to view the hen and other attractive stimuli using the left eye (Andrew and Dharmaretnam, 1991). On days 10 and 11 they also learn to attend to more distant spatial cues, which relies on use of the right hemisphere. To do so they need to experience the disappearance and reappearance of the hen, or some other object on which they have imprinted.

The fact that days 10 and 11 comprise a sensitive period during which the experience of a disappearing/reappearing attachment figure is important was shown by imprinting chicks on yellow tennis balls and then placing each chick in a cage with the ball suspended in the centre and a screen on either side of the ball (Freire *et al.*, 2004; Freire and Rogers, 2005). Opaque screens meant that the chick could move out of sight of the ball. Transparent screens meant that this was not possible. After two or three days in these conditions the chicks were tested on a spatial task designed to reveal whether they attended to local, landmark cues or to distal, spatial cues (i.e. used the functions of the left or right hemisphere, respectively, as shown by Tommasi and Vallortigara, 2004). The chicks that had been given opaque barriers preferred to attend to the distal spatial cues, whereas those given transparent barriers attended to the local, landmark cues. Hence, the experience of going out of sight of the imprinting stimulus, and re-uniting with it, is essential for the development of the right hemisphere's spatial abilities. Facilitated use of spatial abilities during the period of right hemisphere dominance promotes learning to use distal spatial cues to find the imprinting object. The experience must occur during the period when the right hemisphere is dominant, and as little as one day's experience with the opaque barriers on day 11 is sufficient to have the effect, whereas one

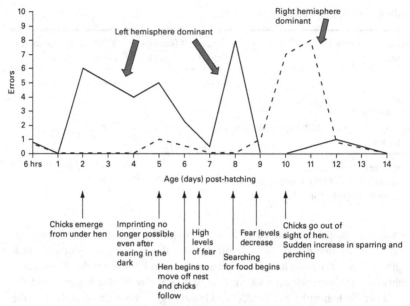

Figure 4.8 Shifts in hemispheric dominance. An illustration of shifts in left and right hemisphere dominance and key steps during the first two weeks of development of the domestic chick. The effect of treating the left hemisphere (continuous line) or the right hemisphere (dashed line) with cycloheximide, which interferes temporarily with neurotransmission and protein synthesis in the injected hemisphere, on learning ability on day 14 is plotted to show which hemisphere is susceptible (and dominant) at the different ages. The measure of learning is the number of pecks at pebbles (errors) in the last 20 pecks of a task in which the chick is allowed 60 pecks to find grains and avoid pecking at pebbles: higher scores show impairment of this ability (Rogers, 1991). Important changes in behaviour are indicated below the x-axis (from Workman and Andrew, 1989; Freire and Rogers, 2007). Note that the latter occur in association with shifts in hemispheric dominance. Adapted from Fig. 13.2 in Rogers (2010b).

day's experience on day 8, when the left hemisphere is dominant, is ineffective (Freire and Rogers, 2007).

Shifts in hemispheric dominance during development are likely to be characteristic of other species too. In mammals, especially humans, this is likely to depend to some extent on development of the corpus callosum, which continues until at least the time of puberty, coupled with differential rates of development of the two hemispheres. Thatcher *et al.* (1987) reported changes in the electroencephalogram activity of the two hemispheres in developing children aged between 3 and 10 years; these changes occurred asynchronously in the left and right hemispheres. It is possible that they might reflect changes in hemispheric dominance as development of the brain takes place.

## 4.6 Developmental effects and lateralization in humans

It is certain that hemispheric asymmetry is present in very young human infants. For example, measurement of hemispheric activity of 3-month-old infants using fMRI has shown left hemispheric dominance in response to hearing speech sounds and many other sounds (Dehaene-Lambertz et al., 2006). Despite some role of the right hemisphere in speech at all ages (Bradshaw, 1989), the left hemisphere appears to have a dominant role in language and speech from very early in postnatal life (Dehaene-Lambertz et al., 2006). This left hemisphere bias for speech seems to depend to some extent on experience (Minagawa-Kawai et al., 2011) and, in addition, the acquisition of speech may affect shifts in laterality of functions other than speech. One example of this may be the apparent shift in hemispheric dominance of categorical perception of colour. Some evidence suggests that, before they can speak, human infants (aged between 4 and 6 months) categorize colours using the right hemisphere, whereas adult humans do so using the left hemisphere and lexical colour codes (Franklin et al., 2008a, 2008b). Perhaps perceiving colour according to categories shifts from the right to left hemisphere as the ability to produce speech develops, but further research on infants of different ages will be required to ascertain the time-course of these changes. Such age-related comparisons are difficult to interpret because it is difficult to test pre-linguistic infants, as Davidoff et al. (2009) have pointed out. One approach to overcome this is to use a variety of tests of colour perception, particularly since, in adult humans, although the left hemisphere is used for categorical, language-based perception of colour, the right hemisphere may be used for non-linguistic and non-categorical perception of colour (Roberson and Hanley, 2009), as well as colour perceived in mental imagery (Howard et al., 1998).

On the other hand, some sort of categorical perception of colour not dependent on knowing the names of colours may well be a function of the right hemisphere in infant humans (Franklin, 2009) and, once names have been associated with particular categories of colour, quite radical changes in functional location may take place. In fact, discrimination of colours from different linguistic categories activates regions of the left hemisphere used for language, as well as regions used in visual perception (Siok et al., 2009). A recent study by Kwok et al. (2011), using magnetic resonance imaging of the brain, has discovered that adults trained to use new names for subcategories of blue and green in just two hours show an increase in the volume of the grey matter in regions V2 and V3 of the left visual cortex. This finding reminds us of plasticity of the brain discussed at the beginning of this chapter. Even in adults, experience can cause changes in brain structure that occur quite rapidly.

Despite these complications in interpreting age-related changes, it is clear that the development of brain and behaviour passes through a series of sensitive periods, at least some of which follow different time-courses in the left and right hemispheres (Rogers, 2010b). Some of the sensitive periods during development are timed precisely but their timing is not independent of experience. The timing of one sensitive period, and what takes place during it, may depend on a previous sensitive period (Rogers, 2010b). As Gilbert Gottlieb (2002) pointed out, development follows trajectories that are not restricted to genetic programmes but include sensory stimulation of the embryo and, we add, postnatal stimulation. Experience-dependent development or learning that takes place during one sensitive period may constrain that which will take place during a later sensitive period.

## 4.7 Conclusion

As a general characteristic of the brain, lateralization is robust but this quality integrates with experience so that brain function adapts to specific environments. It is not a matter of nature versus nurture but of the interplay between the organism and its environment (for further reading on this topic see Bateson and Gluckman, 2011). One could think of it as a picture in which lateralization is sketched in for all, or most, vertebrate species, but in which the details are added flexibly with different influences in different environments. As a result, some processes may or may not be lateralized or may vary in their strength of lateralization depending on how the organism develops and how its plan for development adapts to experiences that inevitably impinge on it.

# Causation

## Summary

Left–right differences are too complex to be summarized by any simple dichotomy. In particular, interaction between the left and right hemispheres is crucial, with shifts in control reflecting collaboration, as well as bringing to bear the specializations of one rather than the other side. A major theme of this chapter is an attempt to show how specializations of one or other hemisphere interact with those of its partner. It has been argued that left hemisphere control is needed for assessment of a stimulus (e.g. assignment to a category) and for the subsequent selection of an appropriate response. However, initial detection of a stimulus is often made by the right hemisphere, owing to its ability to attend widely across panoramic space and to a wide range of properties of the stimulus. The left hemisphere may then intervene to control further assessment. At the same time, when a task is being performed, the left hemisphere is able to specify relevant properties of the stimulus, which are then used in searching for that stimulus using both the right and left hemispheres.

## 5.1 Introduction

Discussions of brain lateralization, understandably, often attempt to make simple summaries of its organization. We might say that we talk with the 'left brain' and are emotional with the 'right brain'. The term 'brain' itself is inadequate here, as we will see when considering the roles of different structures; the term 'hemisphere' (i.e. cerebral hemisphere) is used when it is reasonably clear that forebrain structures are chiefly involved. Such terms are a necessary convenience for broad generalizations. Nevertheless, it is clear that, although shortened characterizations of right and left are unavoidable, at some points in discussions they should be treated with caution. An unexpected recent attempt at such characterization (Dien, 2008) is that the left hemisphere 'anticipates multiple possible futures', while the right hemisphere 'integrates ongoing strands of information into a single view of the past'. Novel dichotomies such as this, however, have the merit of provoking new lines of thought.

It has been argued in previous chapters that left hemisphere control is needed for assessment of a stimulus (e.g. assignment to a category) and for the subsequent selection of an appropriate response. However, initial detection of a stimulus is often made by the right hemisphere, thanks to its ability to attend widely across panoramic space and to a wide range of properties of stimuli. The left hemisphere may then intervene to control further assessment. At the same time, when a searching task is being performed, the left hemisphere is able to specify relevant properties of the stimulus being sought, and these can then be used by the right as well as the left side of the brain.

Intervention by the left hemisphere, immediately following the detection of a stimulus and assessment of its emotional valence by the right hemisphere, is a striking example of hemispheric interaction. By this means, the initiation of an intense emotional response may be held in check, accompanied by a marked reduction in that particular emotion or a change in the assessment of the stimulus itself.

As we said in Chapter 1, in vertebrates with laterally placed eyes visual input crosses the midline, so that the right eye input is directed to the left side of the brain, and the left eye feeds the right. This 'optic decussation' is likely to have been of fundamental importance in the early evolution of our lateralization. Vertebrates to various extents see different things with their left and right eyes. Hearing differs from vision since sounds normally reach both ears and both sides of the brain. Nevertheless there are striking effects on behaviour of blocking one or other ear. In Chapter 3 we gave the example of the lactating female mouse going to the calls of her pups only if her left ear is blocked and her right ear is open (Ehret, 1987). This is apparently because decisions to respond to a stimulus tend to be taken by the left hemisphere, which is especially concerned with stimuli that appear to be located in right-side space. 'Dichotic listening tests' (different sounds played to each ear simultaneously), in which attention of human subjects is directed by prior instruction to inputs played to one or the other ear, or competition is induced by simultaneous competing inputs to the two ears, reveal that sounds perceived to be on the right side are assessed by the left hemisphere and *vice versa*. Auditory input is, therefore, largely attended to by the hemisphere opposite the ear listening for the sound. In this sense, it has some similarity to processing in the visual system.

Left–right differences may be used to constrain how a stimulus is assessed. In birds, the right and left eyes sometimes scan independently with active selection of different stimuli. However, even with the yoked eye movements shown by mammals (Chapter 3), different things can be seen by the right and left visual fields. Throughout the vertebrates, choice of one eye to view a stimulus can be used to control the assessment of specific visual inputs. Such simple exclusive selection of inputs is not possible for hearing since it is not possible for the animal to restrict the input to one ear, but the direction of attention to the right or left can be used to determine which inputs are assessed.

In contrast to vision and audition, olfactory input from the nostrils goes to the cerebral hemisphere on the same side, rather than crossing the midline. Lateralization of the effects of olfactory inputs has been demonstrated (see also Chapter 1). In human infants, presentation of a pleasant food-related smell to the left nostril is more likely to elicit a head turn towards the scent than is presentation of the same odour to the right nostril (Olko and Turkewitz, 2001). Left hemisphere involvement in initiation of response may be responsible for olfactory asymmetry. In dogs, the right nostril is used in initial sniffing of scents that are non-aversive, and a shift to left nostril use follows with repeated presentations (Siniscalchi *et al.*, 2011) (Figure 5.1). No such shift occurs with arousing scents (e.g. the odours associated with veterinary visits). The right hemisphere is thus used to assess scents thoroughly and as long as this is necessary. By choosing a nostril the dog may determine how it will assess a particular stimulus.

The main abilities asymmetrically controlled by the vertebrate brain may be summarized as the ability to select a stimulus for response and to sustain it against distraction, and the ability to attend to and record multiple properties of

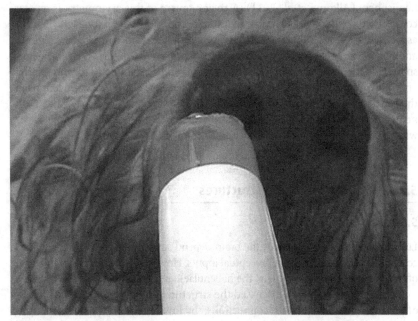

Figure 5.1 Sniffing dog. The test used by Siniscalchi *et al.* (2011) to measure lateralized use of the nostrils. The dog in the photograph is using its right nostril and right hemisphere.

a scene or object (as opposed to the ability to focus attention on a feature selected for response). The first is associated largely with structures on the left side of the brain, and the second largely with structures on the right side. Left hemisphere control also tends to suppress responding to stimuli that evoke 'emotional' responses. These are more readily initiated, and tend to be more intense, with right hemisphere control. Other aspects of lateralization, which are elaborations of the basic specializations of the left side, include left hemisphere control of planned sequences of responses, culminating in handedness and language in humans (Chapter 3).

A measure of overall patterns of brain asymmetry, which has been much used, is cortical thickness. Lüders *et al.* (2006) showed that right-handed adult humans tend to have overall greater thickness of the cortex of the left hemisphere, but noted that variations in asymmetry in this measure are very difficult to interpret. This is due in part to the fact (discussed later) that changes in thickness, which occur during development, are affected by the timing of pruning of connections, as well as by initial numbers of neural cells. In rats the hippocampus is thicker on the left in females and on the right in males (Diamond, 1991) but the functional correlates of such thickness differences are uncertain.

Asymmetries of connectivity are more revealing than the size of brain regions. In humans, the left thalamus has higher connectivity with the left hemisphere, to which it sends perceptual information, than does the right thalamus to the right hemisphere (Alkonyi *et al.*, 2011). A major source of this asymmetry is greater connectivity from the thalamus to the prefrontal areas on the left. This is associated with the left hemisphere being well connected in some specific regions, and playing a role in highly specific and local processes, whereas the right hemisphere is better connected across larger regions and so specialized for integration and more general processes (Iturria-Medina *et al.*, 2011). Neural connections in the left hemisphere have a broad range of conduction speeds, and the right hemisphere has more fast-conducting, myelinated fibres, which allow faster wide-ranging interaction (Nowicka and Tacikowski, 2011).

## 5.2 Asymmetry of brain structures

### 5.2.1 Habenulae

Left and right specializations of the brain depend strongly on forebrain structures concerned with assessment of perceptual inputs. However, asymmetries of function of other regions of the brain (e.g. the habenulae) are also crucial to lateralization.

In Chapters 3 and 4 we discussed the structural and functional asymmetries of the habenulae and their connections. In lower vertebrates the structural asymmetry of this region of the brain is obvious. In mammals no asymmetry of structure is present but the habenulae still perform different functions. It is known in both zebrafish (Aizawa *et al.*, 2005) and mammals (Kappers, 1936, p. 1083) that fibres from the habenulae cross the midline and provide inputs to both sides of the brain. This suggests that, if an emotional input strongly activates the habenula on the right side, the consequences on brain function will be bilateral.

Lateral habenulae are involved in behaviour that requires continuing interaction between the specification of a target and the aim of the behaviour (matters for the left brain), and control of attention (which requires the participation of the right brain). The lateral habenulae, when active, inhibit dopaminergic neurons in the midbrain (Bianco and Wilson, 2009; Hikosaka, 2010). In rats, if this inhibition of dopaminergic neurons is blocked experimentally, the animal is more easily distracted. Thus, when the left side of the brain selects and sustains a response, lateral habenular activation opposes distraction. In zebrafish, genetic silencing of the left lateral habenula causes freezing induced by an electric shock to persist, instead of being progressively replaced by locomotion (Agetsuma *et al.*, 2010). In this case, the 'emotional' input is made more effective by the absence of the left lateral habenula. The choice by zebrafish to use the right eye when approaching food targets (Miklósi and Andrew, 1999; Miklósi *et al.*, 2001) implies control by the left side of the brain and left lateral habenular involvement. This would mean reduced likelihood of distraction.

### 5.2.2 Hippocampi

In mammals, the hippocampus is involved in analysis and linkage of multiple properties of a stimulus, using inputs from a wide range of sensory structures. These inputs have already undergone some perceptual processing. The hippocampi show striking asymmetry at the cellular level in the CA3 pyramidal cells. In rodents ε2 NMDA receptors (a specific type of glutamate receptor) in the synapses of the neurons are more common on apical synapses of CA3 pyramidal fibre terminals in the left hippocampus but on basal synapses in the right hippocampus (Kawakami *et al.*, 2003; Goto *et al.*, 2010). ε2 NMDA synapses develop synaptic plasticity faster than other synapses (Goto *et al.*, 2010), which may affect whether the right or left hemisphere is better in retaining a memory. Learning to select a baited hole out of an array of many holes is slower in mutant rodents in which both sides of the hippocampus have the right-side condition. In a task requiring shifting between levers to obtain a reward, working memory is impaired in such mutants when an interval of more than 10 seconds is required before a response can be made. Hence, encoding longer-term memories depends on having the arrangement of NMDA synapses typical of the left hippocampus.

The hippocampus is important in elaboration of long-term memory records, which in the rat include not only information about a particular place, but also about a specific journey (i.e. place of start and destination), together with records of places along the route (Ferbinteau and Shapiro, 2003; Diba and Buzsáki, 2007). Hippocampal asymmetries are also likely to be related to right hemisphere involvement in analysis of spatial layouts. In humans the right hippocampus is especially involved in short-term memory of the position of objects (Piekema et al., 2006). This agrees well with findings for the domestic chick, in which lesions of the right but not the left hippocampus prevent the use of geometric cues to find the centre of an enclosure, although use of a landmark to find the centre is not affected (Tommasi et al., 2003; Chapter 2).

In humans, the hippocampi are important in assessing relations between multiple properties. This holds for odours, for visual inputs such as faces and for word-colour items with particular associations, as well as for spatial position (Konkel et al., 2008). The hippocampus is also involved in processing novel words, which is accompanied by activation of the left anterior hippocampus (Saykin et al., 1999), as might be expected from the left hemisphere special-izations for verbal performance. Associations between visual and auditory com-ponents of a memory involve the hippocampus, as much as visual–visual associations do (Holdstock et al., 2010). Major reciprocal connections between frontal structures of the brain and the hippocampus probably allow left prefrontal control of the left hippocampus during verbal learning (Shallice et al., 1996); right prefrontal structures are active in memory verification, even though the material retrieved is verbal.

### 5.2.3 Cerebellum

The cerebellum (see Figure 3.6) is chiefly involved in motor responses and coordination. In rodents and primates, it is known that deep cerebellar nuclei predominantly facilitate specific planned responses by the motor cortex (da Guardia et al., 2010). Different parts of the human cerebellum are involved in different types of motor task (Stoodley and Schmahmann, 2009). Since the connections between the cerebrum (cortex) and the cerebellum cross the mid-line of the brain, it was not unexpected to find that, in the cerebellum, the region involved in speech is markedly lateralized to the right and performance on spatial tasks is lateralized to the left (Habas et al., 2009). Lateralization could also be true of the extensive cerebellar connections of the prefrontal 'networks' controlling sustaining and shifting of attention (discussed later).

In chimpanzees, overall cerebellar asymmetry is in opposite directions in the anterior and posterior regions, and the direction of this asymmetry (known as cerebellar torque) is opposite in left- and right-handed individuals (Cantalupo et al., 2008; Cantalupo and Hopkins, 2010). Rightward asymmetry is present in humans in most parts of the cerebellum but not all, and there are also sex differences (Fan et al., 2010). Involvement in advanced motor control, in particular in handedness and speech, is likely to underlie such asymmetries.

In addition, a wide range of other functions (e.g. emotions) is associated with different areas of the cerebellum (Habas *et al.*, 2009).

Cerebellar asymmetries are not confined to primates. Some rays and sharks show them, apparently in association with midbrain asymmetries of unknown function (Puzdrowski and Gruber, 2009).

### 5.2.4 Optic tectum

In vertebrates other than mammals, much of the detection and assessment of targets (e.g. in feeding) is done by midbrain structures, the optic tecta (Chapter 4). These pass on to the forebrain information from the eye by which they are fed, and receive modifying control in return. Much more elaborate classes of information (including pattern and size, as well as colour) are transmitted to the forebrain from the optic tecta in birds than in mammals (Fredes *et al.*, 2010).

In the pigeon, the right and left optic tecta differ in microstructure, and the right eye sends a much greater supply of connections to 'its' forebrain, the left, via the left optic tectum than does the left eye by the corresponding route to the right forebrain (Valencia-Alfonso *et al.*, 2009). As explained in Chapter 4, the left hemisphere receives strong inputs from both eyes, whereas the right hemisphere receives most of its input from the left eye only. Further, the left optic tectum is better able to inhibit its partner on the other side, so that, once inputs from the right eye become the focus of attention, such control will tend to be sustained more readily than would be the case for inputs from the left eye. In addition, since the left forebrain is specialized for recognition of objects being sought, using local features, the left forebrain's greater ability to use information acquired via the optic tecta (Manns and Güntürkün, 2009) should make its control during food search advantageous.

Also, as explained in Chapter 4, a different asymmetry is present in domestic chicks, located in the thalamofugal visual system. The right eye sends inputs via the left side of the thalamus to both hemispheres, whereas the left eye sends its inputs mainly to the right hemisphere (Deng and Rogers, 2002a). It thus seems that in birds there may be asymmetry in one or other of two different routes causing either right or left forebrain structures to get more information from the side of the visual field that is not its main supply. Presumably different functions are associated with each route. In the chick, greater bilateral/binocular supply to the right forebrain is consistent with attentional search throughout the environment for a rare stimulus. The importance of insect prey to fowl may mean that detection of a single target positioned anywhere is crucial. The neural organization of the visual system in pigeons may be a reflection of predominant search, under left forebrain control, for one type of food item (a seed). This food type will commonly be present throughout the environment, across which it is widely scattered and freely available, so that selection must be based on identification of cues defining a specific category (merely of 'seed'), which can be performed best under left hemisphere control.

The mammalian homologue of the optic tectum, the superior colliculus, is still involved in targeting visual stimuli under the control of forebrain visual structures. If damage to these forebrain structures prevents conscious perception, there are still effects of conspicuous stimuli on eye targeting (Mulckhuyse and Theeuwes, 2010). It is possible that lateralization has also persisted in these visual responses of mammals.

## 5.3 Emotion and the amygdala

Much of the evidence relating to the lateralized processing of emotional information concerns the amygdala, a subcortical forebrain structure present in both hemispheres. The importance of the amygdalae in social interaction is shown by the positive correlation between size of the corticobasal amygdalae and size of social network in humans and other primates (Bickart et al., 2011). Also in humans, neurons in the right amygdala respond selectively to living things (see Mormann et al., 2011).

In humans, blocking conscious visual perception by rapid superimposing of a masking stimulus in place of the motivating initial stimulus causes activation of the right amygdala but not the left (Morris et al., 1998). The procedure involved presenting an angry face in association with startling white noise. When, after such training, presentation of the face was immediately masked by subsequent presentation of a neutral face, there was no conscious perception of the angry face, but the appropriate emotion was experienced. At the same time as the conscious perception of the stimulus is blocked, there is experience of the appropriate emotion and the right amygdala is active. Without masking there is conscious perception, and at the same time the left, but not the right amygdala is activated. Assessment by left-side structures therefore follows the initial emotional response by the right-side structures, and allows both conscious perception and the possibility of regulating the levels of emotional response.

The less intense response to emotional stimuli due to left brain control may even be quite different from that resulting from right brain control (e.g. amusement or play rather than fear, attack or disgust: Davidson, 1995). Evidence for this in animals is provided by the evocation of attack more readily by objects seen on the left (and so detected by the right brain) in vertebrates, including toads, lizards, birds and mammals (Chapter 1 and reviewed by Andrew and Rogers, 2002). The hypothesis that negative emotions are associated with activation of the right hemisphere, and positive ones with activation of the left (Davidson, 1995) has been influential. However, positive emotions also tend to be more intense in association with right hemisphere involvement (e.g.

sexual behaviour in humans: Tucker and Frederick, 1989). In the chick, copulation is more easily evoked with the left eye in use (Rogers et al., 1985).

The evidence from 'masking' experiments shows that right brain structures are very effective in the rapid recognition of emotional stimuli, and then further processing brings structures in the left side of the brain in control of responding. The special role of the left hemisphere, including the left amygdala, in controlling the intensity of response to stimuli, which are aversive owing to past experience rather than because of inherent effectiveness, is shown by left amygdalar activation when they are presented (Phelps et al., 2001).

In human subjects, in whom lesions of the primary visual (striate) cortex prevent conscious visual perception, there is still response to emotional facial expressions (de Gelder et al., 2002). When a voice is paired with an emotional facial expression, subsequent assessment of that voice is affected. Mediation by the surviving collicular route to the right amygdala for visual information is thought to be responsible. There is no such effect of pairing with 'emotional' pictures of spiders or puppies, presumably because they require assessment via the visual cortex, which is not available in these subjects.

Left frontal activation in anger is associated in humans with attempts to resolve the agonistic situation (Harmon-Jones et al., 2010). In the chick (Phillips and Youngren, 1986), lesions of the right but not left archistriatum (an amygdalar homologue) depress distress calling, which is a component of distress behaviour. Amygdalar lateralization also affects the ability to maintain a response. Injection of lidocaine (a drug which suppresses neural activity) into the left, but not the right, basolateral amygdala reduced 'hole dipping' in rats in the presence of a distracting novel object (Alvarez and Banzan, 2011). In this case repeated selection of holes, each containing a small amount of food, was sustained only if the left amygdala was available to oppose its partner; presumably, after administering lidocaine to the left amygdala, the balance between left and right forebrain control was shifted away from the left.

## 5.4 Left hemisphere and categorization

One of the most important aspects of lateralization is the way in which the two hemispheres differ in categorization of the properties of objects. Kosslyn et al. (1992) showed that quantitative judgements by humans of the distance between objects are faster in the left visual field (right hemisphere), whereas at least some categorical judgements (e.g. above/below or connected/unconnected) are faster in the right visual field (left hemisphere). Categorization involves deciding where to divide a continuum along which objects vary so as to choose appropriately between such objects.

It was long assumed that assignment to colour categories was largely imposed by the existence of names for colours (the 'Whorfian hypothesis'), and so would be related to verbal abilities of the left hemisphere. This was supported by the finding that, when it was necessary to detect differences between colour samples which might belong to the same or different categories, as defined by colour names used by the subject, there was greater response to differences between categories relative to ones within a category in the right visual field, but not in the left (Gilbert et al., 2006). However, there is also evidence that the perceptual processing of colour space affects the way colour categories are divided in similar ways in many languages (Regier et al., 2007); this suggests that verbal categorization might arise from perceptual processing rather than the other way round (also discussed in Chapter 4).

Before they acquire the names of colours, infants show reverse lateralization compared with that of older individuals. They have greater enhancement of response to between-category differences (shown by latency of eye movement in response to seeing colours) in the left visual field rather than the right (Franklin et al., 2008a; see also Chapter 4). The adult condition, with right visual field enhancement of effectiveness of between-category differences, appears in young children along with the acquisition of colour names (Franklin et al., 2008b). This suggests that the underlying processes are rather like those already described for detection of emotional responses by the right hemisphere, with the possibility of subsequent modulation by the left hemisphere. Right hemisphere mechanisms have initial advantage in analysis of varying perceptual properties, but learning allows the left hemisphere to intervene to select properties that can be used to define a category.

This is also true for stimuli with visual patterning instead of colour as the relevant variable. Adults with full linguistic abilities were asked to allocate novel chequerboard patterns (for which no verbal terms exist) to one of two arbitrary model patterns. Early in learning to do this, right prefrontal structures concerned with control of attention are active (Seger et al., 2000). In later stages of such learning, activation shifts to corresponding left prefrontal structures, and the degree of such activation correlates positively with performance. There are similar shifts even when learning to read or to use a novel script.

The left hemisphere also appears to organize memory of varying properties differently from the right. Differences associated with the use of right or of left hand in response were used to determine this. Thus, when a response has to be given to one number within a specified sequence of numbers, numbers near the end of the sequence evoke faster right hand response, and numbers near the beginning faster left hand response (van Dijck and Fias, 2011). The right hand is also faster for large sizes than small in such a sequence (whether this response is to the size of a stimulus or to a concept of size, such as the words 'table' versus 'mountain'). In the case of a sequence of luminance, the darker end was faster with the right hand, and the lighter end with the left (Ren et al., 2011). These results suggest that the category to which stimuli have to be assigned is defined by the left hemisphere, based on whichever is the clearer end of the range of values forming the category; this end value for numbers is obviously 1. In the case of other sequences such as luminance and size it is less clear *a priori* which this should be; it is perhaps easier to decide that final darkness or very great size has been reached. Regardless of the explanation, when the right hemisphere is in control, it tends to use the end opposite to that used by the left hemisphere.

Once categories have been formed, they will usually be used by the left hemisphere, since assignment to a known category is necessary to guide choice in a world where perceptual inputs vary in unpredictable ways.

## 5.5 Attentional and cognitive mechanisms

Attention and cognition relies strongly on left–right differences; nearly all the evidence derives from human studies. Most data are available for visual inputs, but comparable 'cortical processing streams' exist for hearing (Poremba and Mishkin, 2007). In general terms, 'prefrontal' structures are involved (that is, structures lying behind the forehead). It is possible to distinguish two forebrain networks (dorsal and ventral) controlling attention (in the broad sense), together with a third network known as the 'default' (Corbetta *et al.*, 2008).

The dorsal network, comprised of the anterior prefrontal cortex, dorsal and superior parietal cortex, is activated during the expectation of perceiving a specified stimulus (e.g. at a particular site or with specified properties) and there may be accompanying modulation of activity in the visual cortex. This network produces top-down modulation (explained in Chapter 1) of attention and working memory, when these are involved in current goals and expectations. This involves strong control of the right hemisphere attentional structures of the ventral network (Fox *et al.*, 2006).

Dorsal network functions are clearly related to anterior prefrontal involvement in decision-making (Koechlin and Hyafil, 2007). This includes controlling shifts back and forth between separate tasks, and learning new routines or associations of many serially presented stimuli. Koechlin and Summerfield (2007) argue for an ordered distribution of functions, with the rostral tip of the area arbitrating between responses to several different cues, which may involve controlling shifts between different tasks, and using past learning. When two tasks have to be carried out concurrently, it is the left medial prefrontal cortex that undertakes the primary task (Charron and Koechlin, 2010). Further towards the posterior of the brain, rules about when to perform a chosen response are applied, and further behind that are located effects of context (e.g. changes in circumstance). Finally, in the premotor cortex, sensorimotor associations, as well as the constraints imposed by anterior structures, are involved. Sustaining performance of a task that requires prolonged vigilance is impaired by lateral frontal lesions only when these are on the left (MacPherson *et al.*, 2010), suggesting that such sustaining of response (requiring narrowly focussed attention without distraction) is primarily a function of the left hemisphere.

As said earlier in this chapter, the left thalamus, which relays perceptual information to the left forebrain, has higher cortical connectivity than the right, and a major source of this involves connections to the left prefrontal areas (Alkonyi *et al.*, 2011). This is consistent with the special involvement of the left forebrain in category formation, and categorization is important in choice of response by the left side of the brain. Interestingly, apes show enlargement of the left prefrontal cortex but, in humans, the right prefrontal catches up (Smaers *et al.*, 2011), suggesting that humans may have adapted to associate greater complexity of items of information during thought.

The ventral network involves quite different areas (e.g. supratemporal, supramarginal cortex and medial frontal gyrus). It responds to stimuli that will reorient attention, because they are conspicuous, unexpected or relevant to the task being performed. The dorsal and ventral networks are almost entirely separate in activation, with the exception of some right hemisphere structures associated with the ventral network, such as the temporo-parietal junction and the ventral frontal cortex. These, together with structures like the inferior frontal gyrus, appear to be responsible for the modulation by the dorsal network of the properties being used by the ventral network in searching for a stimulus. Thus the right temporo-parietal junction is activated by stimuli anywhere in space, not only if they are salient but also if they are task-relevant (Fox *et al.*, 2006).

The general control of attention is especially a matter for right hemisphere structures of the ventral network, together with the modulation from the dorsal network of the left hemisphere. Homologues of such ventral network structures in the left hemisphere (areas of Broca and Wernicke) have become central to human speech. Broca's area largely corresponds to the right ventral frontal cortex, and Wernicke's area corresponds to the right temporo-parietal junction. It is likely that the involvement of these areas in language in the left hemisphere relates to control of search amongst internal records, analogous to control of externally directed attention. After selection and association of such records they may be used to control response such as sequences of gestures or of phonemes.

The functions of the right temporo-parietal junction include orienting to and assessing stimuli detected in either visual field (Shulman *et al.*, 2010; Geng and Mangun, 2011). The right temporo-parietal junction has a special role in organizing the body image, as in the 'rubber hand illusion' (Tsakiris, 2010): if a rubber hand can be seen, but not the real hand, and touches to the rubber hand are accompanied by unseen touches to the real hand, the sight of the rubber hand being touched comes to be felt as a touch to the real hand. Disruption of activity in the right temporo-parietal junction opposes acquis- ition of this illusion. A comparable full body illusion can be produced by providing a three-dimensional image of another body and using touches in the same way as for the hand (van der Hoort *et al.*, 2011).

Comparison of the right temporo-parietal junction with Wernicke's area, its homologue on the left, suggests that these two homologous areas are involved on both the left and right in mediating interaction between dorsal and ventral networks. This allows matching of very complex inputs with memories. In the case of Wernicke's area, these are memories of sound patterns corresponding to words, whereas on the right a record of body image is maintained. Overall, left hemisphere structures are dominant in mental imagery of body parts, and the right hemisphere in imagery of the full body, consistent with right hemisphere concern with entire panoramas (Blanke *et al.*, 2010).

In the case of Broca's area, it is likely that the original ability of the right ventral network to search widely for particular types of stimuli has been further elaborated in the human left hemisphere for speech processes (Fox *et al.*, 2006). Lesions of Broca's area produce no deficit when pairs of syllables have to be discriminated if both are heard, but have marked detrimental effects when one syllable is heard and the other seen (Hickok *et al.*, 2011). Activation of Broca's area by brain stimulation facilitates naming pictures; this involves both retrieval of semantic memories and mapping of such retrieved information on to a phonological representation (Holland *et al.*, 2011).

Lateralization is also marked in the execution of 'deductive' tasks, in which the left lateral frontal cortex appears to provide the 'memory space necessary to build the representations created whilst deriving deductive conclusions' (Reverberi *et al.*, 2009). Interestingly, the authors also argue that left lateral lesions, which include the adjacent Broca's area, affect the construction of step-by-step proofs. Each step presumably requires search of available information, which is specified both by the information just acquired and by the overall aim of the task.

Motor control (e.g. hand use) involves the two networks, much as just discussed for speech. In a visuo-motor tracing task, the ventral network is activated, together with structures such as the inferior frontal gyrus that are responsible for interaction between dorsal and ventral networks (Callaert *et al.*, 2011). The involvement of the ventral network was ascribed to the need to continuously exploit spatial information. At the same time, motor structures such as dorsal premotor cortex are activated on the left; this is true even when the left hand is being used by right-handers.

Lesions of the left medial prefrontal cortex appear to promote 'divergent thinking', whereas corresponding damage on the right impairs 'originality' (Shamay-Tsoori *et al.*, 2011). The tests (e.g. list as many alternative uses as possible for a shoe) assessed 'originality' as being reduced by focused attention and increased by 'divergent thinking'. In fact, Chi and Snyder (2011) have found that performance on tasks in which insight is needed is improved by reducing excitability of the anterior temporal lobes in the left hemisphere and increasing excitability of these regions in the right hemisphere, achieved by using electrodes (cathodal stimulation on the left and anodal on the right). This result may be comparable with the Shamay-Tsoori *et al.* (2011) finding, in that it reflects less constraint by established patterns of thinking.

Right frontal structures appear to be involved in assessing the likely organization of possible sequences of responses, rather than directly controlling a planned response. When grasping hand movements are observed, 'mirror neurons' (see Chapter 3) are active, and they are the same ones active when such a hand movement is performed. In the case of Broca's area such units are active in the left hemisphere. However, if the intention of the response of grasping can be inferred by the observer, comparable neurons become active in a structure in the right hemisphere (in the right inferior frontal cortex: Iacoboni *et al.*, 2005). It is perhaps the need to identify the nature of a future action out of a number of possibilities that brings in right hemisphere involvement.

Left hemisphere specializations for control of response have probably resulted in the right hemisphere remaining largely responsible for more general control of attention. Hence, lesions of the right ventral parietal cortex (VPC) tend to produce neglect of visual stimuli (Cabeza *et al.*, 2008); VPC has strong connections with the hippocampus, presumably mediating the association of many items of information. Activity in the right ventral posterior parietal cortex increases during detection of an unexpected stimulus anywhere in the entire visual field (Pisella *et al.*, 2011). This involves appropriate direction of subsequent rapid eye movements, known as saccades. Parietal lesions on the right affecting the temporo-parietal junction (involved in interaction between dorsal and ventral networks, as already discussed) impair such saccades. A corresponding left lesion has no such effect.

Local lesions of the right ventromedial prefrontal cortex produce profound disturbance, in humans, of social behaviour and emotional processing, presumably because of difficulty in detecting and evaluating empathic cues. Corresponding lesions on the left have no such effects (Tranel *et al.*, 2002). Involvement of the right amygdala in emotion has already been discussed. Similar involvement of structures involved in a third forebrain network is considered next.

## 5.6 Emotion and the 'default' network

The effectiveness of stimuli that produce attention shifts due to their emotional significance acts (at least in part) through the cingulate cortex, which is a key component of the 'default network', so called because, as we will see, it is also active when attention is not being directed anywhere in particular in the external environment.

Emotional stimuli (e.g. the face of a loved one, social rejection, pain and empathy for pain) all activate the cingulate cortex. This has attracted recent attention (Seeley *et al.*, 2006; Allman *et al.*, 2011) because of the presence in the human cingulate of von Economo neurons (VENs), which are present also in apes (see Chapter 1). These have been argued to reflect unusual needs in apes and humans for advanced and subtle use of social signals. The presence of VENs in whales and elephants (Semendeferi *et al.*, 2011) perhaps supports this hypothesis, in view of the complex social structures of such animals. However, the same finding is also consistent with association with very large brain size, since VENs are specialized for rapid long-distance integration of information (Craig, 2009; Fan *et al.*, 2011).

VENs are more numerous in the right than the left cingulate cortex and associated structures (e.g. in the insula region: Seeley *et al.*, 2006). Their activity is reduced in autism (low empathy) in the right insula in social but not non-social tasks (Allman *et al.*, 2011). VEN density falls in the right cingulate in early onset schizophrenia (Martin *et al.*, 2010b). The neurons are affected by a very wide range of emotional states: thus they produce neuropeptin signals of satiety, and are affected by wellbeing (e.g. 'sensual touch', warmth), by negative affect (infant cries) and are also involved in risk taking (loss in gambling).

The default network (involving both the cingulate cortex bilaterally and the left inferotemporal cortex) is deactivated during cognitively demanding tasks, and most active while resting with the eyes closed. Its activation is strongly linked to hippocampal activation, together with associated structures, such as the entorhinal cortex and parahippocampal gyrus. These connections presumably allow the assembling of related and relevant information (Greicius *et al.*, 2003, 2004). Activation is particularly strong when there is retrieval of autobiographical memory or envisioning of the future (Spreng *et al.*, 2008). This is associated with 'stimulus-independent' thoughts (Buckner *et al.*, 2008). At the same time the extrastriate visual cortex is deactivated, and the posterior cingulate cortex appears to shift from gathering external information to searching internally (Raichle *et al.*, 2001). The left inferotemporal cortex is perhaps involved as part of the left hemisphere's involvement in deductive tasks (Reverberi *et al.*, 2009); this should be compared with left hemisphere involvement in planning routes.

On the other hand, 'awareness', involving feelings based on the happenings of the immediate moment (and involving both cingulate and VEN), is especially dependent on right hemisphere structures such as the anterior insular cortex (Craig, 2009). It is the right hemisphere that switches at such times between the default system and the other networks. It seems clear that the same lateralization of functions occurs during thinking as it does during assessment of immediate perceptual input.

A possible role of the default network in memory formation is considered in Section 5.12, where it is also shown that the nature of available emotional information is critical to how likely it is that a memory will be retained permanently. This could explain the large involvement in memory formation of structures like the cingulate cortex.

Animals, including humans, sometimes need to search memory independently of direct perception. Default-mode activation occurs in chimpanzees when they are awake but resting (Smaers *et al.*, 2011). In fact, the default system has been demonstrated anatomically in monkeys (Buckner *et al.*, 2008). Fascinating direct evidence for searching through memory in rats is provided by the fact that, at a choice point between two routes in a maze, hippocampal place cells fire in sequences corresponding to the pattern that they had shown during actual passage down first one and then the other route (Diba and Buzsaki, 2007).

## 5.7 Decisions and shifts of assessment

We argue that amusement experienced by humans (and related behaviour in other animals) represents elaborations of mechanisms that control shifts in assessment. As already explained, right hemisphere structures such as the amygdala are especially involved in initiating emotional states, but left hemisphere mechanisms then modulate what follows. The sudden reappraisal of disturbing situations that accompanies amusement and laughter provides the most striking example.

The association of human amusement with predominantly left hemisphere control was early described by Gainotti (1972, 1989), who found that, when patients were informed of their brain damage, those who had left hemispheric damage showed great disturbance and depression, whereas right hemisphere damage was associated with denial of illness and joking. Comparable findings for specifically anterior damage were reported by Robinson *et al.* (1984).

Remarkable effects of control by one or other side of the brain were described by Schiff and Lamon (1989, 1994), who demonstrated that persistent repetition of movements of the right or left hand or the right or left side of the mouth affect mood. Left-side movement produces feelings of sadness (right hemisphere), and right-side movement produces sarcastic or smug moods, which are likely to represent minimization of emotional states (left hemisphere). This phenomenon has been confirmed by Harmon-Jones *et al.* (2010), and linked with asymmetries of prefrontal cortical activation.

### 5.7.1 Laughter

Laughter is a display made by humans given when amused and in friendly play (e.g. being tickled). When tickled by a friendly human of superior status, chimpanzees grin and give breathy sounds, which are broken up into short segments, sounding like the more fully voiced vocalizations given by human children in comparable circumstances (Andrew, 1963). Such 'play panting' is characteristic of the recipient of play attack in wild chimpanzees (Matsusaka, 2004), much as laughter occurs at such times in humans. Short breathy calls also occur in playful wrestling in Barbary macaques (Kipper and Todt, 2002).

Behaviour comparable to laughter has been recognized in rats, so that neither the complex social structures nor elaborate brains of primates are needed for it to be present. Young rats make specific ultrasonic 50 kHz 'chirps' in play fighting or when tickled by a familiar human (Panksepp, 2007). Such behaviour declines in adults, as is usual with play. However, chirps continue to be given to cues associated with a regular meal of desirable food (Panksepp, 2007). Both male and female adults chirp before sexual encounters (McGinnis and Vakulenko, 2003). Presumably chirping in play has a function similar to that of human laughter: namely, reducing the likelihood that vigorous social interaction will escalate into conflict. Dopamine agonists self-administered into the nucleus accumbens also produce chirps in rats (Burgdorf et al., 2001). The accumbens is a subcortical structure, which like the amygdala is involved in motivational states (e.g. the accumbens is activated in food hoarding by gerbils: Hui-Di et al., 2011). In rats, changes in cues that produce 'retrospective revaluation' depend on interaction between accumbens and prefrontal cortex (Saint-Galli et al., 2011).

Laughter is naturally best understood in humans. Most studies of laughter in humans have concentrated on humour and jokes, which are clearly a further and specifically human development of reappraisal of potentially disturbing situations. The sound of laughter is effective in both increasing estimates of funniness and in promoting smiling in the audience (Martin and Gray, 1996). The most important feature of human laughter is its association with reductions in the intensity of potentially unpleasant situations. If such a situation, which is part of a social interaction, can be viewed as amusing, then it is less likely to provoke serious dispute. Smiling and laughing are likely to affect all the participants.

Both the left and right sides of the brain are involved in humour. The right frontal cortex is especially involved in perception of humour, and right frontal lesions produce greater deficits than lesions of the left in ability to distinguish humorous from non-humorous cartoons (Wild et al., 2003). This is consistent with right hemisphere advantage in detecting emotional aspects of perceptual inputs. However, far more regions of the left hemisphere are activated at the same time, including areas involved in reward such as the temporal pole and the supplementary motor area, together with the left amygdala. One specific area, the left fusiform gyrus, is activated at such times, and electrical stimulation of the left fusiform gyrus induces a feeling of mirth (Mobbs et al., 2003). The fusiform gyrus is also activated, along with the amygdala, but on the right side when anorectic people are presented with images of their own body, which is for them an unpleasant experience (Seeger et al., 2002). Once again, homologous structures are involved in emotion, those on the left being associated with mild and pleasant states, and those on the right with intense (and commonly unpleasant) states. The right fusiform region is activated, along with the right amygdala, in response to an unpleasant visual stimulus (Morris et al., 1999);

fusiform activation is here involved in conscious assessment, since it is blocked by perceptual masking. Early signs of assessment of perceptual content occur in the fusiform gyrus on the left in the case of words and on the right for pictures (Maillard *et al.*, 2011). Activation of left brain structures to funny (but not unfunny) cartoons occurs in the left fusiform gyrus, as well as a wider range of structures on the left, including the lateral inferior gyrus (LIG), and Broca's area, which lies within the LIG, and the left amygdala (Mobbs *et al.*, 2003). Finally, right hemisphere damage allows recognition that the ending of a joke is surprising but disturbs any judgement as to which ending is funny (Brownell *et al.*, 1983).

A sex difference exists in identification of facial identity: the left fusiform gyrus tends to be active in women (coupled with more general bilateral activation), but the right fusiform gyrus is active in men (Proverbio *et al.*, 2010). A similar asymmetry has been reported for tests with arousing pictures (Canli *et al.*, 2002). The sex difference described above thus might relate to male attempts to establish identity and female to consider significance.

The left fusiform gyrus is also specifically involved in attempts to establish meaning. When attempting to interpret meaningless artificial Chinese characters, Chinese readers show elevated activation of the left fusiform gyrus, relative to activation during their interpretation of real characters with a meaning (Wang *et al.*, 2011). The left fusiform gyrus is also activated by words in both right and left visual fields, whereas the right is activated only by input from the left visual field (Barca *et al.*, 2011). Importantly, transfer of information from the right to left hemisphere is associated with activation of the left fusiform gyrus, which begins after brief initial analysis by the right hemisphere. The fusiform gyrus is thus associated with wide access to, and use of, relevant material. Remember that left hemisphere structures are involved in the establishment and use of learned categories in assessment.

### 5.7.2 Risk taking

Risk taking resembles amusement in the sense that what might have been distressing is found to be tolerable or even (in humans at least) enjoyable. An evolutionary precursor of this was probably the use of inputs to left hemisphere structures (e.g. from the right eye) to allow sustained inspection of predators and other potential sources of danger, instead of fleeing. This is present in vertebrates from fish onwards (Chapter 3). There is as yet little evidence that the opportunity to do this is sometimes sought after and found to be enjoyable, although elevation in adolescence (of animals) of risk taking and of social play hints that this may be so.

Typically the right eye is used to inspect predators (Bisazza *et al.*, 1997a, 1997b), especially as approach to the predator becomes close (De Santi *et al.*, 2001). Zebrafish use the right eye in a first encounter with a quite novel object

Figure 5.2 Australian magpie, *Gymnorhina tibicen*, looking overhead. The left eye is used to scan overhead for predators. See text for details. Photograph by Professor L. Rogers.

but the left at a second such encounter (Miklósi *et al.*, 1998). The same pattern is present in birds when mobbing predators. Australian magpies use the right eye when approaching a predator but shift to the left eye before withdrawing from it (Koboroff *et al.*, 2008) (Figures 5.2 and 5.3).

Activation of the right prefrontal cortex in humans opposes risk taking: once again the right hemisphere is promoting intense emotional behaviour (cf. in animals, examination of a predator under left hemisphere control). Injury or depression of function of the right prefrontal cortex but not of the left increases risk-taking behaviour, whereas low activity in the right prefrontal cortex correlates with taking larger risks (Gianotti *et al.*, 2009). In rats, exposure to a cat produces long-lasting potentiation of activity in the right amygdala (strongly involved in fear), together with sustained anxiety. The state can be abolished by low-frequency stimulation of the left amygdala (Adamec *et al.*, 2005). The right amygdala activation presumably acts in association with the prefrontal cortex, and it has been argued that high impulsivity in human adolescence relates to delayed functional connectivity of prefrontal structures (Casey *et al.*, 2008).

These changes in the prefrontal cortex are associated with a marked rise in the secretion of androgens by the adrenal glands during adolescence ('adrenarche'). This is especially obvious in chimpanzees and humans (see Chapter 3), where it is accompanied by remodelling of the prefrontal cortex (Spear, 2000), probably affecting risk taking, as well as having more general effects on the control of attention and response by dorsal and ventral networks (as discussed before).

Figure 5.3 Viewing a predator. A taxidermic specimen of a monitor lizard (indicated by the arrow) has been presented to a group of wild apostle birds (*Struthidea cinerea*). Note that the birds nearest to the predator (five birds marked R) are viewing it with their right eye, whereas those slightly further from it are doing so with their left eye (L). This illustrates specialization of the right eye for approach and left eye for withdrawal. Photograph by Professor L. Rogers taken at the field site in far western New South Wales, Australia.

### 5.7.3 Play behaviour

During the well-defined periods of adolescence in mice (days 34–46) and also in rats, there is enhanced risk-taking behaviour in maze exploration. Peaks in social play occur at the same age (Laviola *et al.*, 2003). The role of lateralized mechanisms in play has not been investigated, but it might involve left hemisphere control, which would allow pleasure to be experienced in potentially disturbing situations.

Recognizable play is also present in groups other than apes. In Australian marsupials, juvenile kangaroos show both sudden and unpredictable accelerations in locomotion and, when they are a little older, sparring play occurs (Byers, 1999). Carnivorous Dasyurids perform motor acts of prey capture in play. However, bandicoots, which become independent of their mother within days of birth, appear not to play at all (Byers, 1999). In birds (Ficken, 1977), hawks and eagles show predatory play, repeatedly carrying objects up and then dropping and catching them. Corvids chase in play, and perform elaborate manipulation of objects, sometimes at the same time hanging upside down

(Pellis, 1981) and cockatoos play in very similar ways. Play in other birds is at least not obvious. Reptiles and fish have not been reported to play, with one possible exception: object play with a ball by a freshwater turtle (Burghardt *et al.*, 1996). Hence, there are many possibilities of studying lateralization of play in different species.

## 5.8 Hemispheric interaction and transfer of information

A key feature of advanced mammals is the evolution of the corpus callosum, which connects both corresponding and different areas of the left and right cortices (Schulte and Müller-Oehring, 2010). Initially in evolution, corpus callosum may have no more than supplemented the left–right connections provided by the anterior and tectal commissures in other vertebrates (Figure 4.4). However, the shorter route between the hemispheres, provided by the corpus callosum, extending as it does for much of the length of the dorsal surface of the brain, must have progressively allowed more and more extensive fast interaction between the left and right forebrain.

Transfer of information may reflect temporary access to information in the other hemisphere. The effectiveness of processing of perceptual material, as shown by relative activation (e.g. the right hemisphere tends to be more activated for processing music or facial recognition, but the left for verbal material: Wisniewsky, 1998) also affects transfer. Overall, the hemisphere less competent in assessment (less activated) shows faster transfer to the other hemisphere: thus transfer is faster from right to left hemisphere for verbal material but from left to right for visual material (Nowicka and Tacikowski, 2011). Presumably this allows fast and effective processing to be initially undertaken by the more competent hemisphere. When emission of a response to a visual stimulus is required urgently, there is faster transfer from right to left: the left hemisphere is the one that makes the decision and so should be quickly informed if the right hemisphere detects the stimulus (Putnam *et al.*, 2010; Schulte and Müller-Oehring, 2010). Right hemisphere attention can be biased by the left hemisphere towards cues associated with the intended response (see above, dorsal network).

In dichotic listening tasks, subjects are instructed to direct attention to either the right or left ear during simultaneous verbal inputs to both. People who have had the corpus callosum sectioned analyse right ear inputs accurately, whereas their analysis of left ear inputs is poor. It seems that left ear inputs are normally made available to the left hemisphere via the corpus callosum, and this is confirmed by the fact that, in normal subjects, the larger the corpus callosum, the smaller the difference between left and right ear inputs (i.e. the lower the right ear advantage; Westerhausen and Hugdahl, 2008). Right ear advantage for verbal material is overridden when emotional content of words has to be recognized, which would be expected to engage right hemisphere mechanisms (Voyer *et al.*, 2009).

The effects of degradation of parts of the corpus callosum reveal that the posterior region of the corpus callosum is normally involved in inter-hemispheric collaboration. This region allows interference between conflicting properties to be resolved (e.g. in the Stroop test, where a colour name may be made up of letters of the same colour as the word states or a different colour: Schulte and Müller-Oehring, 2010). When such collaboration between the hemispheres is impaired, ability to choose as instructed (e.g. on colour name but not on the actual colour of the letters) is impaired.

It is likely that there is competition between left and right hemisphere control in searching behaviour. When it is necessary to choose rapidly amongst many similar targets, choice is biased to the left side of the visual display. This holds both for cancellation tasks in humans, and for consuming scattered food grains, as shown in chicks and pigeons (Diekamp *et al.*, 2005; see Chapter 1 and Figure 1.7). Good performance depends on attention over wide areas, for which the right hemisphere should have an advantage. The greater ability of the left hemisphere to sustain response, once a target has been selected, would oppose this.

Studying humans, Johanson *et al.* (2006) have shown that in right-handers the left hemisphere may determine which hand is to carry out an action, allocating it to the left hand when appropriate. When accurate rapid targeting is required the left hand (right hemisphere) may be used because rapid eye movements arrive on target sooner when the left hand is used (Lavrysen *et al.*, 2008). When two tasks have to be undertaken simultaneously, right and left medial frontal cortex may each control one of the tasks (Charron and Koechlin, 2010) (cf. enhanced performance of two tasks simultaneously by the chick, when each task can be allocated to a separate hemisphere; Chapter 2).

## 5.9 Sex differences and hormones

Evidence from the domestic chick links left brain control and the effects of the hormone testosterone (see also the example given in Chapter 4). In the chick, right eye use reduces distraction by novel inedible targets (e.g. pebbles) while the chick is searching for food grains scattered amongst these targets (Mench and Andrew, 1986, and see Chapter 1). A 'search image' of the properties of one type of target is thus more effectively used to sustain search when the right eye (left hemisphere) is used. In tests with both eyes in use, the same ability to sustain search is seen in young chicks treated with testosterone. This is shown by scattering food grains of two colours on a floor and scoring which colour the chick chooses to peck at: testosterone-treated chicks make long runs of pecks at one colour and then the other whereas untreated chicks switch between colours much more often (Andrew, 1972; Andrew and Rogers, 1972; Rogers, 1974). Testosterone treatment also reduces distraction due to the sight of a novel object in chicks trained to run down a runway to obtain food, and makes the chick more likely to return to view the distracting stimulus after it has finished feeding, indicating that it has sustained a memory of the distracting stimulus (Klein and Andrew, 1986). Another effect of testosterone is greater reliance on the frontal visual field: when feeding from moving dishes, testosterone-treated chicks are much less likely than untreated controls to choose to peck at bowls as they enter the lateral visual field, and instead peck at the bowls straight in front of them (Rogers and Andrew, 1989). These findings suggest that testosterone-treated chicks are strongly dependent on the left hemisphere as they search for food. Testosterone may promote the ability of the left hemisphere to sustain use of recently acquired information, as part of its role in keeping to a course of action. This would also contribute to testosterone's effect in elevating levels of attack and copulation. At the same time the left hemisphere of testosterone-treated chicks appears to reduce its inhibition of the right hemisphere, and to elevate aggressive and sexual behaviour (explained in Chapter 4). Testosterone's effect in stabilizing a temporary memory may be used to guide immediate responses and to hold new information for brief periods, later to affect behaviour when opportunity permits.

Effects of sex hormones on brain lateralization may also explain at least some of the differences in behaviour between women and men, although it must be recognized how difficult it is to separate hormonal effects from the effects of experience (for further discussion of this see Rogers, 1999b; Kaplan and Rogers, 2003). In general men perform better than women on feature separation tests, requiring attention to a selected feature and to its relations with others (Robert and Ohlmann, 1994; Collins and Kimura, 1997). In such tests an ability to separate the path of a moving object (an object thrown at a target or caught in movement) from background may be measured. In other similar tests a simple figure has to be found within a larger pattern and, in such tests of 'field independence', judgements of orientation of a feature require other cues to be ignored. In mental rotation tests, which consistently show differences between women and men (Voyer et al., 1995), a memory record of a feature or part of a pattern has to be rotated in the imagination. It is likely that this requires the left hemisphere in order to direct attention to the particular part of the visual input initially analysed by the right hemisphere, and then to suppress further analysis by the right hemisphere. Women are generally more likely to show collaboration between the two hemispheres rather than suppression of the abilities of the right by the goals of the left. This may explain the sex difference on feature separation tasks.

Women perform better than men when it is necessary to remember object identity within an array (pattern assessment). If the spatial layout is unchanged but some pairs of objects are exchanged, women are better at detecting this (James and Kimura, 1997). Comparable female advantage is shown in episodic memory (i.e. memory for a single specific experience). This holds for a wide range of memories: newly acquired facts, the range of different activities carried out in a session, face recognition and verbal tasks (Herlitz *et al.*, 1997). The sex difference is not present at very short retention intervals. Both these examples of female superiority would be explained by more effective use of the abilities of the right hemisphere for accessing memory of patterns made up of multiple items, owing to lesser intervention of the left hemisphere in functioning of the right hemisphere.

The sex hormones may affect lateralization either by acting directly in adulthood, in which case the effects will depend on the levels circulating at the time, or by affecting development. In line with the latter, elevated testosterone during human foetal life is responsible for the greater bulk of right to left hemisphere connections via the isthmus section of the corpus callosum in males than in females (Chura *et al.*, 2010). Probably owing to circulating effects of the sex hormones, performance in feature separation tests by women is least good at the mid-luteal phase of the menstrual cycle (Hampson, 1990). Since oestrogen levels are highest at this time, the effects of oestrogens may be opposite to those of testosterone.

## 5.10 Inter-individual variation and sex differences in lateralization

Interhemispheric connections via the corpus callosum vary between individuals. Much of current evidence concerning variation in lateralization between individuals comes from the use of indices such as handedness or sex as a way of sorting people into groups that show, on the average, differences in lateralization. No such index is ideal.

Humans who do not very consistently use their right hand have more interhemispheric connections manifested both structurally and functionally (Häberling *et al.*, 2011; Nowicka and Tacikowski, 2011). Ambiguous handedness and left-handedness in men is accompanied by less marked reversal of asymmetry in the planum parietale (which correlates with direction of speech and hand control) than in women (Kimura, 1999, pp. 136–137; Foundas *et al.*, 2002; Papadatou *et al.*, 2008). This suggests that left-handedness in men, more commonly than in women, has causes other than full reversal of asymmetry.

If the two hemispheres differ little in their ability to carry out a task, interhemispheric transfer of information will be less necessary. This may be the reason why aphasia (impaired speech and language) is far more likely after left hemisphere damage in men than in women (McGlone, 1980). It may also explain why administration of the sedative-hypnotic drug sodium amobarbital to the left hemisphere by injecting it into the left carotid artery has a much greater effect on speech in men than in women (McGlone, 1980) and why recovery of speech after left hemisphere injury is greater in women than men

(Pizzamiglio and Mammucari, 1985). Greater involvement of both hemi-spheres in language gives better residual ability after injury to one hemisphere.

Direct evidence of more powerful interaction between the hemispheres in women is provided by the fact that stimulation of the motor cortex on one side inhibits activation of muscles of the hand controlled by the other motor cortex and this effect is larger in women than men (de Gennaro et al., 2004). It depends on corpus callosal connections and is consistent with the positive correlation in women, but not men, between size of various divisions of the corpus callosum and performance on a wide range of tests (Davatzikos and Resnick, 1998).

Greater symmetry in women is also reported for reaction to touching of the hand (Huster et al., 2011). In men, touching the right hand produces more stimulation of the mid-cingulate cortex than touching the left hand, but in women there was no difference between the hands. In a task in which there was simultaneous palpation of two nonsense shapes by right and left hands, fol-lowed by an attempt to identify such shapes amongst others, boys performed better with the left hand, whereas girls showed much less difference and could perform the task well with either hand (Witelson, 1976).

Right ear advantage for verbal material is strong in men, but smaller in women (Weekes et al., 1995). This is in agreement with the finding that, in verbal material presented visually, judgements of whether nonsense words rhymed or not involved, in men, the area of Broca but not its homologue in the right hemisphere (Shaywitz et al., 1995). In women both right and left homologous structures tended to be activated.

For spoken words, judgements of rhyming involve unidirectional transfers of information from right to left primary auditory cortex, but at the same time there is reciprocal interaction between the right and left superior temporal gyri (Bitan et al., 2010). Structures on the left are involved in phonological infor-mation, and on the right in speaker voice identity. This interhemispheric connectivity is greater in girls than boys.

Greater activation of the hemisphere more competent in performing a particular task occurs in men (e.g. right hemisphere for music processing and left for verbal tasks; Wisniewsky, 1998). Boys show more right parietal cortex activation in visual face recognition, whereas girls show symmetrical activation (Everhart et al., 2001). Greater interaction between right and left visual inputs in women is shown by their greater ability to fuse Julész patterns, presented separately to right and left eyes, to produce a single composite, stereoscopic pattern (Kimura, 1999, pp. 86–87).

As has already been discussed, in humans the right amygdala responds initially to emotional stimuli, followed by modulation of response by the left. Men and women differ in amygdalar involvement in behaviour. When exposed to an emotional input, higher left amygdala activity correlates with subsequent good recall in women, whereas in men this is true for higher right amygdala activity (Kilpatrick et al., 2006). This suggests stronger modulation of emotion by left hemisphere control under such circumstances in women. Women tend to rate emotional inputs more strongly than men. Since both sexes involve the left rather than the right amygdala when ratings are high (Canli *et al.*, 2002), this might explain the above finding. However, interestingly, there are also marked sex differences when subjects are resting with minimum sensory input (Savic and Lindström, 2008). Women then show greater functional connection (measured by local blood flow and reflecting neural activity that links brain structures) of the left amygdala with its partner, and also with other emotional structures, such as the cingulate and hypothalamus. Men have more connections of the right amygdala with other structures, predominantly frontal lobes and sensorimotor structures. This suggests standing control of emotional response in women even in the absence of motivating inputs, and readiness to recognize emotional inputs in men, perhaps coupled with subsequent assessment and control of response.

Hemispheric connectivity also seems to differ between homosexual and heterosexual men (Witelson *et al.*, 2008) and it may have something to do with the superior performance by homosexual men compared with heterosexual men in verbal fluency tests (Rahman *et al.*, 2003) and superior performance of heterosexual men on feature separation tests (Rahman and Wilson, 2003). However, it is difficult to be certain about such results because they are so dependent on exactly which homosexual men have been selected to test, and they do not control for social class, social oppression and other factors that differ between heterosexual and homosexual men. Social oppression can lead to enhanced verbal ability, for example. The differences between homosexual and heterosexual men are unlikely to be caused solely by hormones acting during development, but may possibly reflect interaction between any such effects and subsequent experience.

## 5.11 Lateralization and disorders of behaviour

Schizophrenia has been linked to disturbances of prefrontal functioning (Barbalet *et al.*, 2011). Normally activity in the caudal left lateral prefrontal cortex is affected by activity in the rostral part of the same structures. This effect is much reduced in schizophrenia, suggesting poor ability to use contextual information (see Section 5.5).

Changes in the ability to recognize emotional stimuli by right brain mechanisms would explain some human behavioural disorders. Psychopaths are deficient in the recognition of emotional cues (especially subtle ones), and this correlates with thinning of frontal and temporal cortex in the right hemisphere, relative to the left, and with reduction in the fasciculus connecting the amygdala with the orbitofrontal cortex on the right but not the left (Hecht, 2011). Inappropriate social behaviour would be made more likely as a result.

Left amygdala pathology correlates with higher anxiety in major depressive disorder (Weniger *et al.*, 2006). In this case, modulation of emotion by the left amygdala appears to have been lessened.

Children who show high-functional autism with no deficits in cooperation or emotional understanding nevertheless have reduced ability to recognize emotional expressions (Downs and Smith, 2004). They thus retain some of the disturbances of empathy characteristic of autism. Autism and Asperger's syndrome have been associated with a variety of disturbances of brain structures much involved in lateralized differences. Pars opercularis, a structure associated on the left with Broca's area, is normally larger on the left, but this is reversed in autism (Tager-Flusberg and Joseph, 2003). This is accompanied by brain enlargement, especially in the posterior region, possibly reflecting reduced pruning of connections between nerve cells during development.

Reduced pruning is associated with reduced top-down control, giving deficits in executive control, in which prefrontal structures are crucial (Frith and Frith, 2003). In Asperger's syndrome there is usually less activation of prefrontal structures in assessing emotional states, and no accompanying amygdalar activation. Autism is also associated with poor use of contextual information, as characterized by difficulty in understanding pretence, irony and deception. In autism, there is difficulty in interpreting complex cues of intent: geometric shapes moving in goal-directed ways are assessed normally when apparently chasing or fighting, but present difficulty when coaxing or tricking (Castelli *et al.*, 2002). Autism also may be accompanied by deficits in planning. This is clear for motor behaviour such as reaching to grasp. Autistic children are slower to initiate hand opening (Mari *et al.*, 2003) and have difficulty in planning series of moves (Booth *et al.*, 2003).

One approach to understanding autism is to see it as an exaggeration of the ability (and inclination) to organize information into systems, particular ones of which become of great interest to particular individuals (Baron-Cohen *et al.*, 2003; Baron-Cohen, 2004). Other evidence suggests that exposure to testosterone during foetal development also increases this ability to systematize and reduces empathy (Knickmeyer *et al.*, 2006). This suggests that there may be increased control by the left hemisphere in both cases. High systemizing might well result from unusually effective and persistent constraint by the left hemisphere of interest and response to a particular category of 'stimuli'. Left hemisphere involvement in categorization has already been discussed. Such constraint would be likely to be accompanied by reduced effectiveness of emotional stimuli ('reduced empathy'), much evidence for which has been given earlier (Section 5.2).

Heightened ability to resist distraction and reduced empathy is likely to be important in social contexts in which high-status individuals benefit from ignoring possible social interactions, while carrying out a course of action. Any potential correlation between strength of lateralization and social status has not yet been investigated.

## 5.12 Memory formation and recall

Studies of the establishment of memory traces in humans frequently concentrate on verbal material. Tulving *et al.* (1994) associated 'encoding' of memory (by which 'sensory input is transformed into an internal representation': Kim, 2011) with special activation of the left prefrontal regions, whereas retrieval of memory is associated with activity of the right prefrontal cortex (Cabeza *et al.*, 2002). Gazzaniga (2000) distinguished between encoding of verbal material and of visual (facial) material, and pointed out that, whilst left hemisphere structures are more involved in the first, the reverse is true of the second. As might be expected, encoding of memory depends heavily on the hemisphere better suited to processing the information involved (Miller *et al.*, 2002) but retrieval of memory is more strongly associated with the right hemisphere, whatever the content of the memory.

Other factors also affect encoding. The left inferior frontal gyrus is initially involved in encoding of verbal, visual or verbal–visual associations (Park and Rugg, 2011). This is followed by hippocampal activation, mainly on the right. This sequence may be due to the left hemisphere being involved in prior selection of items, before the right hemisphere dominates in associating them together. It is probable that recall of specific material from amongst a complexity of different stored information requires multiple cues to be taken into account, and that this involves the specializations of the right hemisphere.

The default network is crucial in recalling memory, which can then be used by the dorsal network (e.g. medial prefrontal cortex) in organizing a new memory (Buckner *et al.*, 2008). The default network activity is disturbed in Alzheimer's disease, as surmised from the concentration of plaques in this region of the brain, perhaps causing memory disturbances (Buckner *et al.*, 2008). The hippocampus also has a role in both encoding and recall of memory. It links records of multiple features involved in an event (Kim, 2011).

Left hemisphere structures are involved in recall of 'time-specific' memories ('episodic': Maguire and Mummery, 1999). Correct selection of material associated in a particular episodic memory seems to require left hemisphere control. The left hippocampus, left parahippocampal gyrus and left temporal structures are all involved.

Memory formation has been most extensively studied in the chick, beginning with the work of Horn (e.g. Horn, 1991). A range of structures are activated on either the right or the left side of the brain at different times during the first few hours of memory formation, as demonstrated by direct recording, lesions and measurements of biochemical changes (reviewed by Johnston and Rose, 2002). Changes in memory recall confirm that lateralized processes are involved (reviewed by Andrew, 2002). Following a brief learning task (association of pecking a novel bead with an unpleasant taste and therefore not pecking it again), there are two series of episodes of brief enhancement of recall, one associated with use of the right eye (and so with left brain structures) and the other with use of the left eye. The series associated with right eye use repeats, starting from the time of training, with a periodicity of 16 minutes and the left eye series with a periodicity of 25 minutes (Andrew, 1991b, 1997). These events are accompanied by points of sensitivity to amnestic agents given to the appropriate cerebral hemisphere, suggesting that they represent brief activation of left or right memory traces. Exactly the same timings are shown by zebrafish (Andrew *et al.*, 2009a).

Pharmacological agents that are known to disrupt memory formation very soon after learning affect the left, but not right, hemisphere (Gibbs *et al.*, 2003), suggesting special activity of the left hemisphere initially, perhaps in association with recording the consequences of the pecking response that was initiated by the left hemisphere.

If the unpleasant taste on the bead is diluted, inhibition of pecking normally disappears at 16 and 25 minutes: that is, first in the left hemisphere and then in the right. Such loss of memory can be prevented by a 'reminder' (the sight of the bead with no opportunity to peck it) at 25 minutes, or by the administration (also at 25 minutes) of a β3 agonist to the right hemisphere (Summers *et al.*, 1996; Anokhin *et al.*, 2002). Right hemisphere processes presumably associated with the unpleasant emotional character of the original experience thus have to be adequately intense at the time of the first right hemisphere event, or memory will be lost. Finally, the overlap of left and right events at 48/50 minutes coincides with the normal establishment of long-term memory.

Here, in a bird, there is initially consolidation on both the left and right sides, and these are at least to some extent independent. Memory formation is, however, subsequently punctuated by points at which inadequacies of one or other trace can cause memory loss. It is likely that interactions occur at these points, which may allow establishment of linkages between right and left records, as well as the possibility of loss.

Memory formation and recall in mammals occurs in the hippocampus and in the basolateral amygdala on both sides, and they take place normally in the unaffected structure, following lesions of the one on the other side (Cammarota *et al.*, 2008). There is thus substantial independence between structures in some aspects of memory formation, and important events in memory formation occur with quite different timings in different structures (Izquierdo *et al.*, 2006). The potential lateralization of these has not yet been investigated.

## 5.13 Conclusion

Left–right interaction in the brain operates over wide timescales. The left hemisphere is able to specify properties of a stimulus to be used in searching controlled by the right hemisphere. Right hemisphere detection of an emotional or salient object is rapidly followed by left hemisphere assessment. If necessary, the left hemisphere will assume control, dismissing further examination by the right hemisphere even if the stimulus is disturbing. The roles played by left–right differences between important structures in the brain, such as the hippocampus, amygdala and prefrontal cortex, are beginning to be understood. Following initial assessment and response, lateralization continues to be important during the consolidation and recall of memory. The left hemisphere's ability to search using specific cues is important in recall of specific experiences, whereas the right hemisphere is better able to retrieve memory using a wider range of information.

# Applications and future directions

## Summary

This chapter considers potential applications of our knowledge about brain lateralization and some of the important questions for future research. First, it covers the potential to improve animal welfare by measuring lateralized behaviour. In doing so, it highlights the role of the right hemisphere in the expression of intense emotions and stress responses and extends this to consider unusual lateralization in humans. Then it outlines some of the areas important for future investigation of lateralization in animals and humans. The latter includes understanding of personality theory, thought processes, and formation and recall of memory. Lateralization in the different sensory systems and the interactions between the hemispheres in natural as well as laboratory contexts are important topics for further investigation.

## 6.1 Application of knowledge of lateralization

Now that lateralization has been established as a general characteristic of the vertebrate brain, it is worthwhile considering how this knowledge might be applied usefully to understanding animal behaviour and what lines of research are likely to be fruitful in the future.

### 6.1.1 Animal welfare

Can knowledge of lateral preferences be usefully applied to improving the welfare of animals? It seems very possible that continuous or exaggerated dependence on the right hemisphere occurs in animals suffering stress and that we might determine which animals are in this condition or which environments induce it by measuring eye, ear or nostril preferences. Although some studies have touched on this question, few have investigated it in any detail.

Minimally invasive techniques of measuring laterality could be applied in a wide range of contexts. In animals with eyes placed laterally, eye preferences to view different stimuli can be determined quite readily, especially as head turning. This reveals which eye/hemisphere is being used to view a stimulus. The latter technique has already been applied to birds (e.g. McKenzie *et al.*, 1998), reptiles (Bonati and Csermely, 2011), horses (Larose *et al.*, 2006; Austin and Rogers, 2012), sheep (Versace *et al.*, 2007) and cows (Robins and Phillips, 2010). Alternatively, lateralized responses of the eyes can be measured by introducing the same stimulus into the left or right monocular field of vision and then comparing the animal's responses on the left and right sides. The latter method has been used to assess fear responses in horses (Austin and Rogers, 2007).

Such measurements can be of particular value in assessing fear behaviour of large animals that can cause serious injuries to humans and other animals. For example, horses use their left eye to view a novel stimulus, and the stronger the left eye bias, the more emotional the horse (Larose *et al.*, 2006). Being viewed by the left eye should signal potential danger to the handler, and this is supported by the finding that, during fights between feral stallions, the attacker views his opponent with his left eye just before he inflicts injury (Austin and Rogers, 2012) (Figure 6.1).

Other controlled experimental procedures can be used to assess lateralization. For example, Siniscalchi *et al.* (2010b) presented dogs with three silhouettes (a dog in a neutral posture, a cat in a defensive-threatening posture and a snake). Each type of stimulus was presented simultaneously on the left and right sides, and head turning was measured. Images of the cat and the snake elicited turning to the left, indicating that, at least initially, the right hemisphere is used to process the information. Reaction times were always shorter when left turns were performed, regardless of the stimulus. These results are consistent with the right hemisphere's role in performing the initial broad scanning and right hemisphere attention to fear-inducing stimuli.

Ear preferences can also reflect which hemisphere is in use. Siniscalchi *et al.* (2008) presented dogs with various sounds via two speakers, one to the dog's left side and one to its right side, while the dog was eating from a bowl midway between the speakers (Figure 6.2). Playing through both speakers various barks of dogs usually elicited turning to the speaker on the dog's right side, indicating a preference for processing in the left hemisphere. By contrast, when the sounds of a thunderstorm were played through the speakers at the same amplitude as the barks, the dogs showed a preference to turn their head to the left side, indicating use of the right hemisphere. Importantly, there was a relationship between head-turning preference and reactivity: the higher the reactivity/fear, the more the turning was to the left side (right

Figure 6.1 Horses attacking. The horse on the left is attacking and has viewed his opponent using his left eye (right hemisphere) before lodging the attack. Image modified from www. zimbio.com/pictures.

hemisphere). In fact dogs that showed unusally strong fear responses, including trembling, panting and taking a long time to return to feeding, were more likely to respond to the barks of other dogs by turning to the speaker on their left side (using their right hemisphere) than were dogs more relaxed when they heard the barks of other dogs. In other words, the ear used reflected the dog's emotional state.

It would now be interesting to apply this method to test the laterality of responses of dogs, and other species, to human voices. If dogs can discriminate between the voices of different humans, or even between different spoken words, this test could be applied to determine which dog–handler combinations are likely to involve use of the left hemisphere, and so aid relaxed performance and an ability to learn and follow rules. By contrast, we predict, dog–handler combinations that make the dog fearful would be indicated by a preference to use the left ear and right hemisphere.

A preference to use the right nostril (and right hemisphere) may also indicate emotional state. As mentioned in previous chapters, dogs use the right nostril to sniff arousing odours, such as the scents associated with visiting a veterinarian and the odour of adrenalin (Siniscalchi *et al.*, 2011,

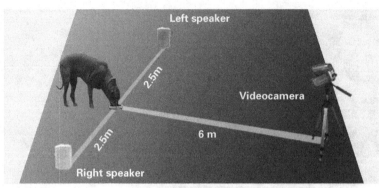

Figure 6.2 Testing lateralized auditory responses in dogs. The sounds (barks or thunderstorms) are played simultaneously through the two speakers when the dog is feeding. The direction of head turning and other behaviour is videotaped. From Siniscalchi *et al.* (2008).

and Chapter 5), and horses use the right nostril to sniff the faeces of an unfamiliar stallion (McGreevy and Rogers, 2005). The effects of such odours on the animal's emotional state have not been assessed directly but should be interesting.

Another way of determining which hemisphere is more active could be to measure lateral bias in the movement of a single midline appendage, such as a tail or trunk. Greater movement to one side of the midline than to the other should reflect greater activity of the hemisphere opposite the side of greater movement (Figure 1.5). Side bias in tail wagging of dogs was mentioned in Chapter 1: greater amplitude of tail wagging to the animal's left side indicates stronger activation of the right hemisphere. Dogs performed tail wagging with a bias to the left side in response to seeing an unfamiliar, dominant dog, and with a bias to the right side when the dog's owner approached (Quaranta *et al.*, 2007). A left side (right hemisphere) bias was also found when the dog was simply standing inside the apparatus (box) used for testing. This may indicate a state of fear or frustration in this condition. Future research measuring the side bias of tail wagging could be a way of assessing individual differences between dogs and emotional responses to a range of stimuli. It is likely, for example, that the approach of some humans would induce fear, whereas others would not. The method could be employed in selecting new owners for dogs in rescue shelters, or more generally, and it would have the advantage of allowing dogs to have a say in choosing a human companion.

Not only does the side bias in tail wagging give us a means of assessing the emotional state of the dog performing the behaviour but also dogs attend to and respond to laterality of tail wagging in conspecifics. This proved to be the case in an experiment in which a robotic model of a dog with a wagging tail was placed in a park where people took their dogs for exercise. Dogs roaming free in the park were more likely to approach the robot, and without hesitation, when its tail was wagging to the left than when it was wagging to the right (Artelle *et al.*, 2010). This means that tail wagging by dogs is a genuine social signal.

Whether tail position to the left or right could be applied to other animals with tails, or other mobile medial organs, remains to be determined. Trunk position might also be assessed in a similar way (e.g. of elephants, Martin and Niemitz, 2003; Haakonsson and Semple, 2009). In long-necked animals, the direction of turning of the neck to position the head on the back during resting may predict behaviour such as aggression, as has been found in flamingos (Anderson *et al.*, 2010): resting rightwards was associated with higher levels of social aggression than was resting leftwards.

Although limited so far, these examples of laterality in wagging or positioning of a medial appendage appear to have promise for assessing fear, aggression and distress in animals. Right hemisphere dominance is of particular concern to questions of animal and human wellbeing (Rogers, 2010a). Persistent control of behaviour by the right hemisphere is not a desirable state. A balance between the two hemispheres is preferable, and may be manifested as an ability to switch from one hemisphere to the other. Sometimes there may be a need for the right hemisphere to assume control (e.g. when emergency responses are required) and at other times for the left hemisphere to do so.

## 6.1.2 Right hemisphere and stress response

In previous chapters we have discussed the role of the right hemisphere in responding to emergency situations. It has a role in both acute and chronic stress responses. A study in mice suggests the principal role of the right hemisphere in processing and modulating pain (Carrasquillo and Gereau, 2008). Two other studies have shown increased blood flow to the right hemisphere in marmosets stressed by being captured (Tomaz *et al.*, 2003) and in cats stressed by being placed in transport containers (Mazzotti and Boere, 2009). These findings are supported by evidence from humans showing that traumatic experience is associated with enhanced activity of the right hemisphere (Shin *et al.*, 1999) and also showing that depression is associated with reduced right-handedness (Denny, 2009) and increased activity of the right hemisphere (Grimm *et al.*, 2008). We can, therefore, deduce that keeping activity of the right hemisphere in balance is in the interests of wellbeing. Via inter hemispheric inhibition the left hemisphere is able to suppress the activity of the right hemisphere (Chapter 5; see also Denenberg, 1981), and a positive effect of this is to reduce the right hemisphere's response to stress (Sullivan, 2004).

The preferred state, it seems, is to preserve a balance between the hemispheres, not in the sense of reducing lateralization but rather by enabling use of either hemisphere as required. No lateralization or weak lateralization would mean that the left hemisphere would be less able to inhibit the right (Rogers, 2010c) and it would also mean impaired ability to perform more than one task simultaneously (discussed in Chapter 2). Strong lateralization with complete dominance of the left hemisphere would diminish an animal's ability to make effective acute responses to stressors (i.e. emergency responses). Complete dominance of the right hemisphere would lead to chronic stress (depression and negative cognitive bias) together with disadvantageous elevation of fear and aggression. To overcome these problems it would be desirable to maintain flexibility of control by either hemisphere in a lateralized brain.

We know that certain conditions during early life can influence the development of lateralization. As discussed in Chapter 4, exposure of developing avian and piscine embryos to light can affect the development of visual lateralization. In rodents, handling in early postnatal life influences the development of brain lateralization and so too does the experience of novelty (Chapter 2, and see Tang and Verstynen, 2002; Tang and Reeb, 2003). In fact, neonatal handling shifts dominance towards the right hemisphere, possibly because it is stressful, and this is not surprising since it involves taking the pups from their mother and isolating them for a certain period each day.

Avoiding stress of various kinds during early life could be a way of preventing right hemisphere dominance. In adult animals already with right hemisphere dominance, and suffering stress, depression and/or increased fear, the solution might be to encourage activity of the left hemisphere. Perhaps this could be achieved by coaxing the animal or human to use the specializations of the left hemisphere, and one way to do this might be to engage the left hemisphere in tasks with simple rules. Following of simple rules consistently, as happens in play behaviour, might achieve the desired effect.

### 6.1.3 Limb preference as a correlate of behaviour

We have discussed the potential of using side bias of moving a medial organ to indicate which hemisphere is being used. In species that use their forelimbs to manipulate objects or parts of the animal's own body, limb preference may indicate which hemisphere is being used. Since each limb is controlled by its opposite hemisphere, consistent limb preferences in individuals may indicate a preferencc to use the hemisphere opposite the preferred limb (Rogers, 2009). Some tasks demand use of the specializations of a particular hemisphere: for example, difficult spatial tasks demand use of the right hemisphere and, hence, as shown in primates, preferred use of the left hand. However, simple tasks can be performed readily by either the left or right hand/hemisphere (e.g. in primates, picking up pieces of food from the floor or an open bowl). Limb preference shown in such a simple task may reflect an individual's preference to use a particular hemisphere, and hence to express a particular set of behaviour.

In Chapter 3, differences in boldness and emotionality associated with hand preference in primates were discussed (Section 3.3). Common marmosets (*Callithrix jacchus*) show individual hand preferences to pick up food and, provided that they perform the task in a relaxed state, about half of them are left-handed and half right-handed (Hook and Rogers, 2000; Rogers, 2009). These individual hand preferences are consistent across a number of reaching tasks (Hook and Rogers, 2008) and across the marmoset's lifetime (Hook and Rogers, 2000; Rogers, 2009). The important point relevant to animal welfare is that limb preferences measured in this way reflect consistent differences between individuals in their propensity to suffer fear or be stressed.

Left-handed marmosets are more fearful of novel environments (Cameron and Rogers, 1999) and fear-inducing stimuli (Gordon and Rogers, 2010) than are right-handed marmosets. The same has been shown in Geoffroy's marmosets (*Callithrix geoffroyi*): they are less likely to sniff a novel food and, when they hear the call of a predator, left-handers freeze for longer than right-handers (Braccini and Caine, 2009). Right-handed marmosets appear to be more responsive to their social group, in that being with social companions increases their likelihood of capturing unfamiliar insects and they use calls to solicit group members to participate in mobbing more often than do left-handed marmosets (Gordon and Rogers, 2010). Right-handed chimpanzees are also more likely than left-handed ones to play with novel toys (Hopkins and Bennett, 1994). Hence, hand preference in simple reaching reflects the behaviour expressed and probably does so because it reflects which hemisphere is dominant in processing perceptual information.

In addition, hand preference may indicate cognitive bias (i.e. decision about whether an ambiguous stimulus is negative or positive; see Bateson and Matheson, 2007 Mendl *et al.*, 2009). A yet unpublished experiment by Gordon and Rogers (in preparation) tested common marmosets that had been trained to differentiate between bowls with a black or a white lid, only one of which was rewarded. Once training had been completed, the marmosets were given probe trials using a bowl with a grey lid. Left-handed marmosets were significantly more likely to treat the grey lid as indicating the negative, unrewarded stimulus than were right-handed marmosets. Left-handers were more reluctant to approach the bowl with the grey lid and they were less likely to remove the lid to look inside the bowl.

Hence, in primates, because limb preference determined on simple reaching or holding tasks is a useful indicator of individual differences in behaviour and cognitive style, it may have a number of uses for improving welfare. Whereas it need not be disadvantageous to react strongly to novel stimuli in the short term, repeated exposure to novel stimuli and environments can, in the long term, cause chronic levels of stress (Kaplan *et al.*, 2012) and, via long-term elevation of stress hormone levels, cause detrimental physiological changes. Therefore, knowledge of a primate's hand preference may be useful in choosing appropriate handling and housing conditions.

Limb preference to manipulate objects has also been measured in dogs, using a number of methods including the paw used to wipe a piece of tape from the nose (Quaranta *et al.*, 2004) and the paw used to stabilize a container while licking food from it (Wells, 2003; Branson and Rogers, 2006). One study has shown that dogs with weaker (non-significant) paw preferences are more likely to suffer from noise phobia, in that they show strong stress responses to certain sounds like thunderstorms, than are dogs with strong (left or right) paw preference (Branson and Rogers, 2006). Dogs with no significant paw preference also stand out from both left- and right-pawed dogs by producing higher levels of one of the stress hormones, adrenalin, some eight days after vaccination for rabies (Siniscalchi *et al.*, 2010a).

These findings about dogs and paw preference are consistent with reports of weaker hand preference in humans suffering from post-traumatic stress dysfunction (Parker *et al.*, 1999) and a number of other cognitive disturbances (e.g. Grimm *et al.*, 2008; see also McGilchrist, 2009). In humans, some of these conditions are associated not only with weaker hand preference but also with left hand preference (right hemisphere dominance) (Shin *et al.*, 1999), whereas this does not seem to be the case in dogs. Nevertheless, since the sample size of dogs tested was quite small, further research on potential association of limb preference and personality in dogs needs to be carried out.

Limb preference in animals that do not use their forelimbs to pick up food or manipulate objects can also be measured, but deduction of which hemisphere is in control is more complicated. Austin and Rogers (2007) measured the direction in which horses turned when they were approached from the front by a person opening an umbrella. They found that horses that turned to the right took flight to a greater distance than did horses that turned to the left. Right turning seems to require propulsion achieved by the left hindlimb and, if so, it probably indicates right hemisphere use in more fearful horses. Alternatively, as Austin and Rogers (2007) argued, right turning would allow use of the left eye during the turn and this would also indicate a preference to use the right hemisphere.

Zucca *et al.* (2011a) determined forelimb preferences in lions using a method that had been developed to measure forelimb preference in horses (McGreevy and Rogers, 2005): that is, scoring the forelimb placed in front of the other during grazing in the case of horses or just standing in the case of the lions. The limb preference of lions showing illness was compared with that of healthy lions and it was found to be significantly stronger in the ailing animals. Both groups had a preference to put the right forelimb in front but this bias was stronger in the sick lions. The authors interpreted the stronger right forelimb preferences of the sick lions as reflecting greater use of the left hemisphere and, therefore, at odds with the prediction that they should use the right hemisphere owing to the stress of illness. However, since the limb placed forward may not be the one supporting most of the body's weight, it may not be the one to consider. The centre of gravity is in the forequarters of the lion, over its forelimbs, control of which would be critical for balance. Ailing lions, it seems, have stronger need to use the left forelimb to support their weight and this would reflect use of the right hemisphere.

Limb preference might also change in strength and/or direction in different housing conditions, as Zucca et al. (2011b) have found in donkeys. Using the standard measure of forelimb preference (i.e. which foot is held in advance of the other), they showed that donkeys changed their preferences when they were moved from a larger to a smaller enclosure. In this case, the move weakened the strength of a right limb preference.

These studies show that direction and/or strength of limb preference is associated with health status and possibly social stress. Limb preference could be a potentially valuable measure to assess the level of stress/suffering and cognitive state of captive animals, and even of animals in their natural habitat.

### 6.1.4 Understanding unusual functioning of the human brain

Some associations between hand preference and mental/emotional state in humans have already been mentioned in previous chapters and earlier in this chapter. Disturbances of mental function are increasingly being shown to relate to unusual lateralization. Often seen as reduced right-handedness, unusual lateralization has been associated with psychosis (Chapman and Chapman, 1987), schizotypy/schizophrenia (Crow, 1997; Shaw et al., 2001; Berlim et al., 2003), severe depression (Grimm et al., 2008; Denny, 2009), post-traumatic stress disorder (Spivak et al., 1998; Zeitlin et al., 1989) and alexithymia, referring to difficulties in understanding and describing emotions (Parker et al., 1999). Schizophrenic patients show impaired prefrontal connectivity when maintaining working memory (Esslinger et al., 2011); they fail to uncouple the right dorsolateral prefrontal cortex from the left hippocampus. This effect was found to be specific to tasks that required working memory. Reduced connectivity via the genu region of the corpus callosum associated with schizophrenia could possibly be involved. Typically the hippocampus is deactivated during working memory tasks. In schizophrenia there may be aberrant recruitment of associated memories, and possibly this is a source of delusions.

From brain lesions caused by stroke and from images of brain activity, a number of studies have associated higher activity of the frontal lobe of the right hemisphere with depressive symptoms (i.e. lesion or hypofunction of the left anterior lobe). In contrast, higher activity of the left frontal lobe (i.e. lesion or hypofunction of the right anterior lobe) is associated with uncalled-for cheerfulness (summarized in Chapter 2 of McGilchrist, 2009; see also page 138).

Functions of the right hemisphere, unchecked by left hemispheric inhibition, are also important in savants. Autistic savants are unusual in their ability to attend to details and attend to literal information. Many can accurately assess large numbers of objects at a glance. According to Allan Snyder and colleagues, autistic savants have privileged access to lower levels of sensory information (bottom-up processing; Snyder and Mitchell, 1999; Snyder *et al.*, 2006). Normal subjects make hypotheses and follow rules that filter their perception of the world. Autistic savants are less governed by rules and categorize objects to a far lesser extent (see Vallortigara *et al.*, 2008). The differences between autistic savants and control subjects strongly suggest that the left hemisphere is less influential in the savants.

Some people with advancing senile dementia pass through a stage when they express abilities to reproduce detailed images and skills that they had never shown before and they exhibit increased interest and ability in painting (Miller *et al.*, 1998). This raises the possibility that the creativity and other functions of their right hemisphere have been released from left hemisphere inhibition. Exactly what functional changes are responsible remains obscure, but it seems clear that the steps in producing a painting based on what can be seen would provide excellent material for studying the interaction between right and left sides of the brain during a creative process.

## 6.2 Future research

It is clear that the techniques outlined in this book will allow lateralization to play a larger part in studies of animal behaviour. At the same time, high-level processes associated with the forebrain dorsal, ventral and default networks (Chapter 5) are likely to illuminate the ability of animals to think, in ways separate from responding to immediate perceptual inputs. The evolution of these abilities is almost completely unstudied.

Equally, measurements largely developed to study animals, such as patterns of eye use (head turning), may be used in studies of humans because they do not require expensive apparatus and are readily applicable to entirely natural contexts.

### 6.2.1 Lateralization in animals: directions of research

So far most, although not all, of the examples of lateralization in non-human animals have been obtained in tests conducted in controlled laboratory conditions or in captive zoo or domestic animals. Some examples of directional biases have been observed in wild animals, and these are important. They include the eye used by kookaburras to watch for moving prey (Rogers, 2002b), eye preferences in stilts when fishing and side bias in copulation (Ventolini et al., 2005), eye preferences in magpies for viewing a predator and catching food (Hoffman et al., 2006; Rogers and Kaplan, 2006; Koboroff et al., 2008), in Caspian terns for angling of prey during feeding of young (Grace and Craig, 2008), eye preference before attack in feral horses (Austin and Rogers, 2012), eye preference to view novel stimuli in dolphins (Siniscalchi et al., 2012), turning biases in tadpoles (Rogers, 2002c) and eye preferences to view prey and a predator by lizards (Bonati et al., 2008; Martin et al., 2010a). More detailed observations of lateralization in each species performing different tasks would be valuable.

In particular, side biases in social displays would be interesting because not only would they reveal more about the expression of laterality in natural conditions but also the direction of the laterality expressed might reveal more about what is being communicated by the display. In horses, the left eye is used preferentially before attacking (Figure 6.1), but displays with less intense emotions, especially highly ritualized ones, may reveal a right eye preference.

In previous chapters (especially Chapter 2) we have covered the hypothesized association between population biases and social behaviour (see also Vallortigara and Rogers, 2005). One might also expect differences in strength or direction of lateralization depending on the social position of individuals within a species. One example of this has been reported in chicks: more strongly lateralized chicks acquire a higher position in the social hierarchy than do less lateralized ones (Rogers and Workman, 1989). Variations in lateralization might be important to take into account when studying lateralization in free-living/wild species.

So far we have not mentioned lateralization of the magnetic compass in birds. Migrating birds use the Earth's geomagnetic field as a navigation guide and do so using two mechanisms (for details see review by Wiltschko and Wiltschko, 2005). In their beaks these birds have magnetite and receptors that detect the magnetic field. In their right eye they have a chemical compass that relies on a photoreceptor called cryptochrome which aligns itself in the ambient magnetic field to which it responds with different spin states of its electrons (Wiltschko et al., 2011). The right eye bias for the magnetic compass was shown first in European robins (Wiltschko et al., 2002) and then in silvereyes (Wiltschko et al., 2003) and domestic chickens (Rogers et al., 2008). In chickens, the right eye superiority in orienting using the magnetic compass seemed to result from the lack of focused attention on magnetic cues when using the left eye rather than an inability to detect the magnetic field using the left eye. This is an important distinction and deserves further investigation using different tasks and testing for lateralization of the magnetic compass in other migratory species known to use a magnetic compass (for examples see Wiltschko and Wiltschko, 2009). The ability of the left hemisphere to sustain long-distance orientation using other senses (discussed in Chapter 3) is consistent with the specialized role of the left hemisphere in orientation using the magnetic field.

By far the majority of examples of lateralization in animals are for the visual modality. We have just mentioned the lateralization of magnetic orientation in birds and there are some but very few examples of auditory, olfactory and tactile lateralization. Future research could fruitfully fill the gap by investigating potential lateralization in non-visual modalities and by looking at their development. Once more evidence of lateralization in other modalities has been obtained, it will be important to consider interactions between the modalities. One such example concerns the interaction between vision and olfaction in the chick (Rogers *et al.*, 1998). Chicks were presented with a small, coloured bead together with a noxious odour. When the bead was blue, the chicks attended to the odour, as shown by shaking of the head, regardless of which nostril they were using, but when the bead was a more attractive red colour, they attended to the odour only when they used their left nostril/left hemisphere, not when they used their right nostril/right hemisphere. Colour so distracted the right hemisphere's attention that response to the odour did not occur. The left hemisphere's ability to categorize and/or to sustain attention may be the explanation why this hemisphere could attend to both colour and odour.

### 6.2.2 Lateralization in humans: directions of research

Indices such as handedness have provided a vast body of data on lateralization in humans (see Corballis, 2002; McManus, 2002). It is now time to begin to interpret and study further existing findings and theories on lateralization in humans. In fact, much needs to be done to understand the basics of human lateralization. The ability to categorize objects, animals or events affects performance in a variety of tasks. In its simplest aspect, categorization requires decisions to be made about appropriate division points in continuously varying properties. This needs to be done in order to make the best choices of responding. The left hemisphere anchors its decision points to one end of the variation (Ren *et al.*, 2011; Chapter 5), whereas the right hemisphere tends to use the other end. Examples of this were given in Chapter 5 (Section 5.4), one being speed of response to a number in a specified sequence, in which case the right hand was faster for numbers near the end of the sequence and the left hand for numbers near the beginning. It will be fascinating to establish what underlies this difference.

Lateralization is just beginning to illuminate personality theory (DeYoung *et al.*, 2010). At least three of the 'Big Five' personality variables are unambiguously associated with asymmetries of function. For example, high Neuroticism (sensitivity to negative emotions) is associated with reduced volume in the left hemisphere of temporal structures, including the hippocampus. One possible interpretation of this asymmetry is that this affects the ability to assess causes of uncertainty and anxiety, which involves the hippocampus because of its ability to associate many pieces of information. Another possible explanation is that the temporal lobe in the right hemisphere has a greater role than normal and hence the heightened sensitivity to negative emotions. High Agreeableness (concern for the needs and desires of others) is associated with reduced length of the left superior temporal sulcus. The structures affected are involved in the interpretation of the actions of others. Perhaps this enhances agreeableness because it reduces interpretation of the actions of others as potentially conflicting with the needs of the observer. High Conscientiousness (ability to constrain oneself so as to follow rules or pursue non-immediate goals) is associated with enlargement of the left medial frontal gyrus. In this case the left hemisphere's ability to sustain response, once initiated, is almost certainly involved (Chapter 5).

A very large body of information on individual differences in behaviour is thus now open to studies based on measures of lateralized brain function. For example, psychopaths (Chapter 5) show deficiencies in recognition of emotional cues, and this is coupled with reduced thickness of the frontal cortex in the right hemisphere and reduced connection of the right of amygdala with the orbitofrontal cortex.

Applications of these types of findings to therapy will no doubt eventually become important. Assessment of risk of contracting certain diagnosable conditions is already possible (e.g. on anatomical grounds). It remains to be seen when treatments such as reducing right hemisphere activity to reduce severe and chronic depression may become useful.

### 6.2.3 Fundamental features of lateralization

It is popular but misleading to make simple characterizations of lateralization (e.g. left hemisphere for language and right hemisphere for creativity). Also, it should be kept in mind that complementary specializations of the hemispheres are not absolute. Nevertheless it is important to distinguish the fundamental differences between the sides of the brain. From the now numerous studies of lateralization in animals we can conclude that the left eye/right hemisphere is used for detecting and responding to predators or opponents of the same species and right eye/left hemisphere for detecting and responding to prey or other food items. This complementary specialization has been shown in fish (Miklósi et al., 1998), toads (Vallortigara et al., 1998), lizards (Bonati and Csermely, 2011) and birds (Vallortigara et al., 2001; summarized by Rogers, 2008). So far, in mammals, there is evidence of use of the left eye/right hemisphere for responding to predators or inflicting aggression on conspecifics (dunnarts, Lippolis *et al.*, 2005; horses, Austin and Rogers, 2012; baboons, Casperd and Dunbar, 1996) but there has been no specific demonstration of use of the left hemisphere for capturing prey or obtaining other food items.

The fundamental difference in function between the hemispheres has been characterized as right hemisphere for emergency responses and left hemisphere for following established rules in familiar, non-arousing circumstances (MacNeilage *et al.*, 2009). In Chapter 5 (Section 5.7), we mentioned the possible role of the right hemisphere in opposing risk taking, which would be necessary in emergency situations, and the role of the left hemisphere in categorization (Section 5.4) and in selecting a goal and sustaining response towards it. These latter specializations would fit with the emergency/routine dichotomy.

Along with the realization that having an asymmetrical brain is not unique to humans, new questions have arisen about higher cognitive functions and their evolution. First, we have been surprised by the fundamental similarities between lateralization in the various vertebrate species studied so far and hemispheric specialization in humans. These fundamental similarities can be considered as evidence that the hemispheric specializations of humans are elaborations on a plan already present in the first vertebrates.

Not surprisingly, since lateralization was thought to be unique to humans, we know more about lateralization in the human species than in other species. In humans, lateralization is more complex and modified in a number of ways from the fundamental pattern delineated in animals. However, the more non-human species are studied, the more we recognize the complexity of their cognitive abilities, and the more similarity to humans we find in their brain lateralization and interhemispheric communication. In non-human animals, as in humans, lateralization is evident in communication and tool using (MacNeilage *et al.*, 2009). It did not appear *de novo* in humans.

In his comprehensive book, Iain McGilchrist (2009) details the differences between the hemispheres of humans. As he suggests, the hemispheres make individually coherent but incompatible representations of the external world and are in tension with each other. We interpret the latter as referring to the importance of interaction between the hemispheres (see below). McGilchrist says that left hemisphere is focused and abstracting. It understands explicit information and deals with it in discrete packages, including discontinuities in time. The left hemisphere is disengaged from context and, consequently, carries out its functions impersonally and without empathy. By contrast, the right hemisphere deals with implicit information and with the whole picture in context. In fact, the right hemisphere deals with the individual's concept of self in context and realistically, as opposed to the left hemisphere's predilection to self-aggrandizement and confabulation. The right hemisphere is given to understanding others and so to have empathy and be cooperative. As found also in many species, the right hemisphere of humans responds to new events and stimuli, which would also be an aspect of its ability to relate to objects and events in context. Together with these attributes, the right hemisphere has the ability to appreciate music and metaphor, and its activity is associated with depression.

How many of these more complex left–right specializations are shared with non-human apes and other species? With the development of tests to measure empathy in animals (e.g. in rats, Langford *et al.*, 2006) it will be possible to determine lateralization of these more elaborate functions of the hemispheres. Not only will such investigations tell us more about brain function in animals but also they will provide an accurate means of experimentally testing some of the ideas about lateralization of human behaviour.

### 6.2.4 Interactions between the hemispheres

Although the differing specializations of right and left hemispheres, or sides of the brain, are fundamental to understanding brain function, the interactions between them are just as important and often quite neglected. Briefly, and as we have discussed in detail in Chapter 5, the role of the left hemisphere in the selection of a particular response and carrying it out requires the left side of the brain to control the attentional abilities of the right side of the brain. Initially the right hemisphere allows wide search (of the environment and/or the memory) specified by multiple properties. Then the left side of the brain prevents shifts of attention (i.e. prevents distraction), which could happen if the right hemisphere detects other (arousing/emotional) stimuli. Of course, there must be rapid and effective shifts of attention when really high priority stimuli are detected. Otherwise animals would be completely vulnerable to predation while they are engaged, for example, in feeding. At any one time, effective performance requires interaction between the left and right hemispheres (i.e. at all times both sides can be or are involved).

In humans, there is evidence that the left hemisphere suppresses the right hemisphere's attention to detail, divergent thinking and insight. Chi and Snyder (2011) have shown that suppression of activity of the left temporal lobe while simultaneously enhancing activity of the right temporal lobe enhances insightful solution of a task. In other experiments, Allan Snyder and colleagues have used repetitive transcranial magnetic stimulation (rTMS), in which a magnetic device is placed on the left anterior temporal lobe, to remove inhibition of the right anterior temporal lobe by the left. During the period of stimulation, the 'freed' right hemisphere allows the subjects to attend to details that they would normally overlook. For example, rTMS increased the accuracy of guessing the number of items (50 to150) that had been shown to them briefly as an array in an image (Snyder *et al.*, 2006). Unconsciously they had attended to the detail of the number of elements and, as Snyder *et al.* (2006) argue, they were temporarily in a mental state approaching that of a savant (see above). Such savant-like abilities, these researchers suggest, are common to us all but they are normally held in check by the left hemisphere.

Inhibition of the left anterior temporal lobe with rTMS also reduces measured scores of prejudice, whereas inhibition of the same area on the right side has no effect (Gallate *et al.*, 2011). During stimulation the test subjects were given a test in which they could associate words related to 'terrorist' and 'law-abider' with 'good' or 'bad' and 'Arab' and 'Non-Arab'. Of course, what is being tested by these associations depends entirely on the particular society and the learned attitudes of people. Prejudiced associations were found to be lower when the left side was stimulated, meaning that neural activity was higher on the right side. The implication of these results is that stereotyped views of social groups are held more strongly in the left hemisphere, and that less prejudiced attitudes are typical of the right hemisphere. The result is consistent with the specialization of the left hemisphere to follow established patterns and, in this case, the rules of the particular society. It is also consistent with the right hemisphere having empathy.

This recent research illustrates the wider implications of our understanding of lateralization on human thought like creativity and insight.

### 6.2.5 Displaying and concealing facial expressions

The evolutionary framework for the study of lateralization outlined in Chapter 2, based on theory of games and the concept of evolutionarily stable strategies (ESS), may provide novel insights even to the more classical approach focused on higher cognitive processing in humans and other animals. A case in point that could be related to brain asymmetry is face processing. There is evidence that portraits, especially since the Renaissance, have been and are typically produced with the left side of the face over-represented, with the head turned slightly to the sitter's right (Hall, 2008) (Figure 6.3). The leftward bias seems to be determined by the sitters and their desire to display the left side of their face, which is controlled by the emotive, right cerebral hemisphere (Nicholls *et al.*, 1999). Thus, whereas the left cheek would be more emotionally expressive, the right cheek would be more impassive. Therefore, the appropriate cheek to put forward would depend on the circumstances of social interaction and, of course, any conscious intent need not be involved at all.

It has been suggested that the motivation to portray emotion, or conceal it, might explain why portraits of males show a reduced leftward bias and also why portraits of scientists from the Royal Society show no leftward bias (Nicholls *et al.*, 1999). Semi-wild orang-utans have been found to preferentially expose the left side of their face to observers, as they look sideways at the observer (Kaplan and Rogers, 2002). This posture of sideways looking would also mean that the right hemisphere would be used.

These strategies of showing the face would be effective only assuming a population structure for the lateral biases, and are thus potentially open to

Figure 6.3 Portrait of a gentleman. This is a typical example of featuring the left side of the face. This portrait of a Venetian gentleman was painted jointly by Giorgione and Titian, painters who lived in the fifteenth to sixteenth centuries.

mathematical modelling similarly to the case of prey-predator interactions studied by Ghirlanda and Vallortigara (2004). Consider expressivity of the face. On the one hand, there could be advantages in terms of predictability of behaviour if, in all individuals, the left side of the face is more 'expressive'. You could approach conspecifics on a particular side and avoid attacks. Yet we know that animal communication has not evolved to transmit honest information but rather (in many cases) to deceive (Krebs and Dawkins, 1984). From this point of view, having an unpredictable (50:50) expressive side of the face may confer advantages. However, if all individuals choose unpredictability (i.e. to conceal emotion) then a disadvantage for every single individual would arise. Thus the expected outcome would be a majority of individuals that behave cooperatively (aligned) and a minority (frequency-dependent, meaning that their number cannot increase too much because their individual advantage is maintained only if they remain a minority and so are only rarely encountered by cooperative individuals) not aligned.

The classical approach to the study of lateralization based on investigation of causal mechanisms, and thus linked to behavioural biology and neurobiology, may certainly benefit from integration with a functional approach, linked to evolutionary biology.

### 6.2.6 Role of lateralization in memory

Lateralization plays a key role in memory formation, which is only just beginning to be studied. It is perhaps obvious, once the issue is considered at all, that since the right and left sides of the brain receive different perceptual inputs, problems of interrelating what has been recorded are likely to arise. This is shown to be the case by evidence from insects and nematodes (Chapters 2 and 3). In fruitflies, the absence of an unpaired, asymmetrical structure in the forebrain in certain mutants prevents the formation of long-term memory. Presumably one side of the brain usually takes the lead in organizing the memory record and, if this is disturbed, no usable trace is established. In bees, there is clear evidence of lateralized memory formation and recall (Chapter 1): a population bias has now been shown in four species of bee to use the right antenna for recall of short-term memory and the left antenna for recall of long-term memory (Rogers and Vallortigara, 2008; Frasnelli et al., 2011).

Evidence (Chapter 5) suggests that, in birds, interaction between material held on left and on right sides may occur at precisely timed points of activation of the memory trace. Evidence from the zebrafish for similar points of trace activation may indicate that this mechanism is common amongst vertebrates but it remains to be seen whether anything similar occurs in mammals.

## 6.3 Conclusion

Research on lateralization has taken us far from the original notion that it was a characteristic found only in the human brain and that it evolved with language and tool using. The basic plan of lateralization was present in the first vertebrates and, it now appears, even in some invertebrates. Although lateralization in invertebrate brains may well have a different origin and depend on different genetic and experiential influences than it does in vertebrates, some of the advantages of having a divided brain may be achieved regardless of brain size and complexity. Besides, when coordinating the asymmetric behaviour of individual animals, invertebrates and vertebrates may have had similar selective pressures to align their asymmetries at the population level. Benefits of being lateralized are superior ability to perform more than one task at a time and aided recall of long- and short-term memories.

Perhaps the issues of greatest general interest lie in understanding the processes underlying thought. Now that the default system, which is central to thought carried out independently of incoming perceptual inputs, has been identified anatomically in humans, its functioning can be studied in animals. The first evidence for comparable processes in animals is becoming available. Such studies will certainly have implications not only for scientists but also for philosophers.

# References

Abe, K. & Watanabe, D. (2011). Songbirds possess the spontaneous ability to discriminate syntactic rules. *Nature Neuroscience*, **14**: 1067–1074.

Adamec, R. E. & Morgan, H. D. (1994). The effect of kindling of different nuclei in the left and right amygdala in the rat. *Physiology and Behavior*, **55**: 1–12.

Adamec, R. E., Blundell, J. & Burton, P. (2003). Phosphorylated cyclic AMP response element binding protein expression induced in the periaqueductal gray by predator stress: Its relationship to the stress experience, behavior and limbic neural plasticity. *Progress in Neuro-psychophysiology*, **27**: 1243–1267.

Adamec, R. E., Blundell, J. & Burton, P. (2005). Neural circuit changes mediating lasting brain and behavioural response to predator stress. *Neuroscience and Biobehavioral Reviews*, **29**: 1225–1241.

Adelstein, A. & Crowne, D. P. (1991). Visuospatial asymmetries and interocular transfer in the split-brain rat. *Behavioral Neuroscience*, **105**: 459–469.

Ades, C. & Ramirez, E. N. (2002). Asymmetry of leg use during prey handling in the spider *Scytodes globulosa* (Scytodidae). *Journal of Insect Behaviour*, **15**: 563–570.

Adret, P. & Rogers, L. J. (1989). Sex difference in the visual projections of young chicks: A quantitative study of the thalamofugal pathway. *Brain Research*, **478**: 59–73.

Agetsuma, M., Aizawa, H., Aoki, T. *et al.* (2010). The habenula is crucial for experience-dependent modification of fear responses in zebrafish. *Nature Neuroscience*, **13**: 1354–1356.

Aizawa, H., Bianco, I. H., Hamaoka, T. *et al.* (2005). Laterotopic representation of left–right information onto the dorsoventral axis of a zebrafish midbrain target nucleus. *Current Biology*, **15**: 238–243.

Aizawa, H., Goto, M., Sato, T. *et al.* (2007). Temporally regulated asymmetric neurogenesis causes left–right difference in the zebrafish habenular structures. *Developmental Cell*, **12**: 87–98.

Albert, M. (1973) A simple test of visual neglect. *Neurology*, **23**: 658–664.

Aljuhanay, A., Milne, E., Burl, E. *et al.* (2010). Asymmetry in face processing during childhood measured with chimeric faces. *Laterality*, **15**: 439–450.

Alkonyi, B., Juhász, C. A., Muzik, O. *et al.* (2011). Thalamocortical connectivity in healthy children: Symmetries and robust developmental changes between 8 and 17 years. *American Journal of Neuroradiology*, **32**: 962–969.

Allan, S. E. & Suthers, R. A. (1994). Lateralisation and motor stereotypy of song production in the brown headed cowbird. *Journal of Neurobiology*, **25**: 1154–1166.

Allman, J. M., Tetreault, N. A., Hakeem, A. & Park, S. (2011). The von Economo neurons in apes and humans. *American Journal of Human Biology*, **23**: 5–21.

Almécija, S., Moyà-Solà, S. & Alba, D. M. (2010). Early origin of human-like precision grasping: A comparative study of pollical distal phalanges in fossil hominins. *PLoS One*, **5**(7): e1727.

Alonso, Y. (1988). Lateralization of visual guided behavior during feeding in zebra finches (*Taeniopygia guttata*). *Behavioural Processes*, **43**: 257–263.

Alonso, J., Castellano, A. & Rodriguez, M (1991). Behavioral lateralization in rats: Prenatal stress effects on sex differences. *Brain Research*, **539**: 45–50.

Alvararez, E. O. & Banzan, A. M. (2011). Functional lateralisation of the baso-lateral amygdala neural circuits modulating the motivated exploratory behaviour in rats: Role of histamine. *Behavioral Brain Research*, **218**: 158–164.

Alves, C., Chichery, R., Boal, J. G. & Dickel, L. (2007). Orientation in the cuttlefish *Sepia officinalis*: Response versus place learning. *Animal Cognition*, **10**: 29–36.

Alves, C., Guibé, M., Romagny, S. & Dickel, L. (2009). Behavioral lateralization and brain asymmetry in cuttlefish: An ontogenetic study. *Poster at XXXI International Ethological Conference, Rennes, France.*

Anderson, M. J., Williams, S. A. & Bono, A. (2010). Preferred neck-resting position predicts aggression in Caribbean flamingos (*Phoenicopterus rubber*). *Laterality*, **15**: 629–638.

Andrew, R. J. (1963). The origin and evolution of the calls and facial expressions of the primates. *Behaviour*, **20**: 1–109.

Andrew, R. J. (1972). Changes in search behaviour in male and female chicks, following different doses of testosterone. *Animal Behaviour*, **20**: 741–750.

Andrew, R. J. (1976). Use of formants in the grunts of baboons and other non-human primates. *Annals of the New York Academy of Science*, **280**: 673–693.

Andrew, R. J. (1983). Lateralisation of emotional and cognitive function in higher vertebrates, with special reference to the domestic chick. In: J. P. Ewert, R. R. Capranica & D. J. Ingle (eds.), *Advances in Vertebrate Neuroethology*, Oxford: Oxford University Press, pp. 477–510.

Andrew, R. J. (1991a). The nature of behavioural lateralization in the chick. In: R. J. Andrew (ed.), *Neural and Behavioural Plasticity. The Use of the Chick as a Model*, Oxford: Oxford University Press, pp. 536–554.

Andrew, R. J. (1991b). Cyclicity in memory formation. In: R. J. Andrew (ed.), *Neural Plasticity. The Use of the Domestic Chick as a Model*. Oxford: Oxford University Press, pp. 476–506.

Andrew, R. J. (1997). Left and right hemisphere memory traces: Their formation and fate. *Laterality*, **2**: 179–198.

Andrew, R. J. (1999). The differential roles of right and left sides of the brain in memory formation. *Behavioral Brain Research*, **98**: 289–295.

Andrew, R. J. (2002a). Memory formation and lateralisation. In: L. J. Rogers & R. J. Andrew (eds.), *Comparative Vertebrate Lateralisation*, Cambridge: Cambridge University Press, pp. 582–633.

Andrew, R. J. (2002b). The earliest origins and subsequent evolution of lateralisation. In: L. J. Rogers & R. J. Andrew (eds.), *Comparative Vertebrate Lateralisation*, Cambridge: Cambridge University Press, pp. 70–93.

Andrew, R. J. & Dharmaretnam, M. (1991). A timetable of development. In: R. J. Andrew (ed.), *Neural and Behavioural Plasticity: The Use of the Domestic Chicken as a Model*, Oxford: Oxford University Press, pp. 166–176.

Andrew, R. J. & Rogers, L. J. (1972). Testosterone, search behaviour and persistence. *Nature*, **237**: 343–356.

Andrew, R. J. & Rogers, L. J. (2002). The nature of lateralisation in tetrapods. In: L. J. Rogers & R. J. Andrew (eds.), *Comparative Vertebrate Lateralization*, Cambridge: Cambridge University Press, pp. 94–125.

Andrew, R. J. & Watkins, J. A. S. (2002). Evidence for cerebral lateralisation from senses other than vision. In: L. J. Rogers & R. J. Andrew (eds.), *Comparative Vertebrate Lateralization*, Cambridge: Cambridge University Press, pp. 365–382.

Andrew, R. J., Dharmaretnam, M., Györi, B. *et al.* (2009a). Precise endogenous control of right and left visual structures in assessment by zebrafish. *Behavioural Brain Research*, **196**: 99–105.

Andrew, R. J., Mench, J. & Rainey, C. (1982). Right–left asymmetry of response to visual stimuli in the domestic chick. In: D. J. Ingle, M. A. Goodale & R. J. Mansfield (eds.), *Analysis of Visual Behavior*, Cambridge, MA: MIT Press, pp. 225–236.

Andrew, R. J., Osorio, D. & Budaev, S. (2009b). Light during embryonic development modulates patterns of lateralisation strongly and similarly in both zebrafish and chick. *Philosophical Transactions of the Royal Society of London B*, **364**: 983–989.

Andrew, R. J., Tommasi, L. & Ford, N. (2000). Motor control by vision and the evolution of cerebral lateralisation. *Brain and Language*, **73**: 220–235.

Anfora, G., Frasnelli, E., Maccagnani, B. *et al.* (2010). Behavioural and electrophysiological lateralisation in a social (*Apis mellifera*) but not a non-social (*Osmia cornuta*) species of bee. *Behavioural Brain Research*, **206**: 236–239.

Anfora, G., Rigosi, E., Frasnelli, E. *et al.* (2011). Lateralization in the invertebrate brain: Left–right asymmetry of olfaction in bumble bee, *Bombus terrestris*. *PLoS One*, **6**: e18903.

Annett, M. (2002). *Handedness and Brain Asymmetry: The Right Shift Theory*. Hove, UK: Psychology Press.

Annett, M. (2006). The distribution of handedness in chimpanzees: Estimating right shift in Hopkins' sample. *Laterality*, **11**: 101–109.

Anokhin, K. V., Tiunova, A. A. & Rose, S. P. R. (2002). Reminder effects – reconsolidation or retrieval deficit? Pharmacological dissection with protein synthesis inhibition following reminder for a passive-avoidance task in young chicks. *European Journal of Neuroscience*, **15**: 1759–1765.

Artelle, K. A., Dumoulin, L. K. & Reimchen, T. E. (2010). Behavioural responses of dogs to asymmetrical tail wagging of a robotic dog replica. *Laterality*, **16**: 129–135.

Asami, T., Gitternberger, E. & Falkner, G. (2008). Whole-body enantiomorphy and maternal inheritance of chiral reversal in the pond snail *Lymnaea stagnalis*. *Journal of Heredity*, **99**: 552–557.

Austin, N. P. & Rogers, L. J. (2007). Asymmetry of flight and escape turning responses in horses. *Laterality*, **12**: 464–474.

Austin, N. A. & Rogers, L. J. (2012). Limb preferences and lateralization of aggression, reactivity and vigilance in feral horses (*Equus caballus*). *Animal Behaviour*, **83**: 239–247.

Babcock, L. E. & Robison, R. A. (1989). Preferences of Palaeozoic predators. *Nature*, 337: 695–696.

Baguñà, J., Martinez, P., Paps, J. *et al.* (2008). Back in time: A new systematic proposal for the Bilateria. *Philosophical Transactions of the Royal Society of London B*, 363: 1481–1491.

Baltin, S. (1969). Zur Biologie und Ethologie des Talegalla-Huhns (*Alectura lathami* Gray) unter besonderer Berücksichtigung des Verhaltens wärrend der Brutperiode. *Zeitschrift für Tierpsychologie*, 6: 524–572.

Balzeau, A. & Gilissen, E. (2010). Endocranial shape asymmetries in *Pan paniscus, Pan troglodytes* and *Gorilla gorilla*, assessed via skull based landmark analysis. *Journal of Human Evolution*, 59: 54–69.

Bambach, R. K., Bush, A. M. & Erwin, D. H. (2007). Autecology and the filling of eco-space: Key metazoan radiations. *Palaeontology*, 50: 1–22.

Barbalet, G., Chambers, V., Domenich, J. D. P. *et al.* (2011). Impaired hierarchical control within the lateral prefrontal cortex in schizophrenia. *Biological Psychiatry*, 70: 73–80.

Barca, L., Cornelissen, P., Simpson, M. *et al.* (2011). The neural basis of the right visual field advantage in reading: An MEG analysis using virtual electrodes. *Brain and Language*, 118: 53–71.

Baron-Cohen, S. (2004). *The Essential Difference*. London: Penguin Books.

Baron-Cohen, S., Richler, J., Bisarya, D. *et al.* (2003). The systemizing quotient: An investigation of adults with Asperger syndrome or high-functioning autism with normal sex differences. *Philosophical Transactions of the Royal Society of London B*, 358: 361–374.

Bateson, M. & Matheson, S. M. (2007). Performance on a categorisation task suggests that removal of environmental enrichment induces 'pessimism' in captive European starlings (*Sturnus vulgaris*). *Animal Welfare*, 16: 33–36.

Bateson, P. & Gluckman, P. (2011). *Plasticity, Robustness, Development and Evolution*. Cambridge: Cambridge University Press.

Beaumont, J. G. (ed.) (1982). *Divided Visual Field Studies of Cerebral Organization*. London: Academic Press.

Bekoff, A. & Kauer, J. A. (1884). Neural control of hatching: Fate of the pattern generator for the leg movements of hatching in post-hatching chicks. *Journal of Neuroscience*, 4: 2659–2666.

Berezinskaja, T. L. & Malakhov, W. (1995). The fine structure of the eye of *Serratosagitta pseudoserratodentata*. *Zoologichesky Zhurnal*, 74: 129–133.

Berlim, M. T., Mattevi, B. S., Belmonte-de-Abreu, P. & Crow, T. J. (2003). The etiology of schizophrenia and the origin of language: Overview of a theory. *Comprehensive Psychiatry*, 44: 7–14.

Berrebi, A. S., Fitch, R. H., Ralphe, D. L. *et al.* (1988). Corpus callosum: Region-specific effects of sex, early experience, and age. *Brain Research*, 438: 216–224.

Bertram, B. (1970). The vocal behaviour of the Indian Hill Mynah *Gracula religiosa*. *Animal Behaviour Monographs*, 3: 79–192.

Bianco, I. H. & Wilson, S. W. (2009). The habenular nuclei: a conserved asymmetric relay station in the vertebrate brain. *Philosophical Transactions of the Royal Society of London B*, 364: 1005–1020.

Bickart, K. C., Wright, C. I., Dantoff, R. J. *et al.* (2011). Amygdala volume and social network size in humans. *Nature Neuroscience*, 14: 163–164.

Billiard, S., Faurie, C. & Raymond, M. (2005). Maintenance of handedness polymorphism in humans: A frequency-dependent selection model. *Journal of Theoretical Biology*, **235**: 85–93.

Binkofski, F. & Buccino, G. (2004). Motor functions of the Broca's region. *Brain and Language*, **89**: 362–369.

Bisazza, A., Cantalupo, C., Robins, A., Rogers, L. J. & Vallortigara, G. (1996a). Pawedness in toads. *Nature*, **379**: 408.

Bisazza, A., De Santi, A., Bonso, S. & Sovrano, V. A. (2002). Frogs and toads in front of a mirror: Lateralisation of response to social stimuli in tadpoles of five anuran species. *Behavioural Brain Research*, **134**: 417–424.

Bisazza, A., De Santi, A. & Vallortigara, G. (1999). Laterality and cooperation: Mosquitofish move closer to a predator when the companion is on their left side. *Animal Behaviour*, **57**: 1145–1149.

Bisazza, A., Facchin, L. & Vallortigara, G. (2000). Heritability of lateralization in fish: Concordance of right–left asymmetry between parents and offspring. *Neuropsychologia*, **38**: 907–912.

Bisazza, A., Lippolis, G. & Vallortigara, G. (2001). Lateralization of ventral fins use during object exploration in the blue gourami (*Trichogaster trichopterus*). *Physiology and Behavior*, **72**: 575–578.

Bisazza, A., Pignatti, R. & Vallortigara, G. (1997a). Detour tasks reveal task- and stimulus-specific neural lateralisation in mosquitofish (*Gambusia holbrooki*). *Behavioural Brain Research*, **89**: 237–242.

Bisazza, A., Pignatti, R. & Vallortigara, G. (1997b). Laterality in detour behaviour: Interspecific variation in poeciliid fish. *Animal Behaviour*, **54**: 1273–1281.

Bisazza, A., Rogers, L. J. & Vallortigara, G. (1998). The origins of cerebral asymmetry: A review of evidence of behavioural and brain lateralization in fishes, amphibians, and reptiles. *Neuroscience and Biobehavioral Reviews*, **22**: 411–426.

Bitan, T., Lifshitz, A., Breznitz, Z. *et al.* (2010). Bidirectional connectivity between hemispheres occurs at multiple levels in language processing but depends on sex. *Journal of Neuroscience*, **30**: 11576–11585.

Blanchard, B. A., Riley, E. P. & Hannigan, J. H. (1987). Deficits on a spatial navigation task following prenatal exposure to ethanol. *Neurotoxicology and Teratology*, **9**: 253–258.

Blanke, O., Ionta, S. & Fornari, E. (2010). Mental imagery for full and upper human bodies: Common right hemisphere activation and distinct extrastriate activations. *Brain Topography*, **23**: 321–332.

Boleda, R. M., Chincilla M., Valls, R. & Pastor, J. (1975). El dextrismo en el chimpancé. *Zoo*, **23**: 18–20.

Boles, D. B., Barth, J. M. & Merrill, E. C. (2008). Asymmetry and performance: Toward a neurodevelopmental theory. *Brain and Cognition*, **66**: 124–139.

Bonati, B. & Csermely, D. (2011). Complementary lateralisation in the exploratory and predatory behaviour of the common wall lizard (*Podarcis muralis*). *Laterality*, **16**: 462–470.

Bonati, B., Csermely, D. & Romani, R. (2008). Lateralisation in the predatory behaviour of the common wall lizard (*Podarcis muralis*). *Behavioural Processes*, **79**: 171–174.

Bone, Q. (1972). *The Origin of Chordates*. Oxford: Oxford University Press, pp. 6–10.

Bonetti, C. & Surace, E. M. (2010). Mouse embryonic retina delivers on formation controlling cortical neurogenesis. *PloS One*, **5**(12): e15211.

Booker, R. & Quinn, W. G. (1981). Conditioning of leg position in normal and mutant Drosophila. *Proceedings of the National Academy of Sciences USA*, **78**: 3940–3944.

Boorman, C. J. & Shimeld, S. M. (2002). The evolution of left–right asymmetry in chordates. *BioEssays*, **24**: 1004–1011.

Booth, R., Charlton, R., Hughes, C. *et al.* (2003). Disentangling weak coherence and executive dysfunction: Planning drawing in autism and attention-deficit disorder. *Philosophical Transactions of the Royal Society of London B*, **358**: 387–392.

Borod, J. C., Koff, E., Perlman, P., Lorch, M. & Nicholas, M. (1986). The expression and perception of facial emotion in brain-damaged patients. *Neuropsychologia*, **24**: 169–180.

Boughman, J. W. (1998). Vocal learning by greater spear-nosed bats. *Proceedings of the Royal Society of London B*, **265**: 227–233.

Boycott, A. E. & Diver, C. (1923). On the inheritance of sinistrality in *Lymnaea peregra*. *Proceedings of the Royal Society of London B*, **95**: 207–213.

Braccini, S. & Caine, N. G. (2009). Hand preference predicts reactions to novel foods and predators in marmosets (*Callithrix geoffroyi*). *Journal of Comparative Psychology*, **123**: 18–25.

Bradshaw, J. L. (1989). *Hemispheric Specialization and Psychological Function*. Chichester: John Wiley and Sons.

Bradshaw, J. L. (1991). Methods for studying human laterality. *Neuromethods*, **17**: 225–280.

Bradshaw, J. L. & Nettleton, N. C. (1982). Language lateralization to the dominant hemisphere: Tool use, gesture and language in hominid evolution. *Current Psychology*, **2**: 171–192.

Bradshaw, J. L. & Rogers, L. J. (1993). *The Evolution of Lateral Asymmetries, Language, Tool Use and Intellect*. San Diego, CA: Academic Press.

Brain, W. R. (1941). Visual disorientation with special reference to lesions of the right hemisphere. *Brain*, **64**: 224–272.

Braitenberg, V. (1984). *Vehicles: Experiments in Synthetic Psychology*. Cambridge, MA: MIT Press.

Braitenberg, V. & Kemali, M. (1970). Exceptions to bilateral symmetry in the epithalamus of lower vertebrates. *Journal of Comparative Neurology*, **138**: 137–146.

Branson, N. J. & Rogers, L. J. (2006). Relationship between paw preference strength and noise phobia in *Canis familiaris*. *Journal of Comparative Psychology*, **120**: 176–183.

Breedlove, S. M., Watson, N. V. & Rosenzweig, M. R. (2010). *Biological Psychology: An Introduction to Behavioral, Cognitive, and Clinical Neuroscience*. 6th edn. Sunderland, MA: Sinauer Associates, Inc.

Broad, K. D., Mimmack, M. L. & Kendrick, K. M. (2000). Is right hemisphere specialization for face discrimination specific to humans? *European Journal of Neuroscience*, **12**: 731–741.

Broca, P. (1865). Sur le siège de la faculté du langage articulé. *Bulletin de la Société d'Anthropologie de Paris*, **6**: 377–393.

Brooks, R., Bussière, L. F., Jennions, M. D. & Hunt, J. (2004). Sinister strategies succeed at the cricket World Cup. *Proceedings of the Royal Society of London B, Biology Letters*, **271**: S64–S66.

Brown, C., Gardner, C. & Braithwaite, V. R. (2004). Population variation in lateralized eye use in the poeciliid *Brachyraphis episcopi*. *Proceedings of the Royal Society of London B*, **271**: S455–S457.

Brown, C., Western, J. A. C. & Braithwaite, V. R. (2007). The influence of early experience on, and inheritance of, cerebral lateralization. *Animal Behaviour*, **74**: 231–238.

Brownell, H. H., Michel, D., Powelson, J. *et al.* (1983). Surprise but not coherence; sensitivity to verbal humour in right-hemisphere patients. *Brain and Language*, **18**: 20–27.

Buckner, R. L., Andrews-Hanna, J. R. & Schacter, D. L. (2008). The brain's default network – anatomy, function and relevance to disease. *Annals of the New York Academy of Science*, **1124**: 1–39.

Budaev, S. & Andrew, R. J. (2009a). Shyness and behavioural asymmetries in larval zebrafish (*Brachydanio rerio*) incubated in the dark. *Behavior*, **146**: 1037–1052.

Budaev, S. & Andrew, R. J. (2009b). Patterns of early embryonic light exposure determine behavioural asymmetries in zebrafish: A habenular hypothesis. *Behavioural Brain Research*, **200**: 91–94.

Budil, P., Thomas, A. T. & Harbinger, F. (2008). Exoskeletal architecture, hypostomal morphology and mode of life of Silurian and Lower Devonian dalmantid trilobites. *Bulletin of the Geosciences*, **83**: 1–10.

Burgdorf, J., Knutson, B., Panksepp, J. *et al.* (2001). Nucleus accumbens amphetamine microinjections unconditionally elicit 50 kHz vocalisations in rats. *Behavioral Neuroscience*, **115**: 940–944.

Burghardt, G. M., Ward, B. & Rosscoe, R. (1996). Problem of reptile play: Environmental enrichment and play behaviour in a captive Nile soft-shelled turtle, *Trionyx trjunguis*. *Zoo Biology*, **15**: 223–238.

Byers, J. A. (1999). The distribution of play among Australian marsupials. *Journal of Zoology London*, **247**: 349–356.

Byrne, R. W. & Byrne, J. M. E. (1991). Hand preferences in the skilled gathering task of mountain gorillas (*Gorilla g. berengei*). *Cortex*, **27**: 521–546.

Byrne, R. A., Kuba, M. J. & Griebel, U. (2002). Lateral asymmetry of eye use in *Octopus vulgaris*. *Animal Behaviour*, **64**: 461–468.

Byrne, R. A., Kuba, M. J. & Meisel, D. V. (2004). Lateralised eye use in *Octopus vulgaris* shows antisymmetric distribution. *Animal Behaviour*, **68**: 1107–1114.

Cabeza, R., Ciaramelli, E., Olson, I. R. *et al.* (2008). The parietal cortex and episodic cortex: An attentional account. *Nature Reviews Neuroscience*, **9**: 613–625.

Cabeza, R., Dolcos, F., Graham, R. *et al.* (2002). Similarities and differences in the neural correlates of episodic memory, retrieval and working memory. *NeuroImage*, **16**: 317–330.

Callaert, D. V., Vercantrien, K., Peters, R. *et al.* (2011). Hemispheric asymmetries of motor versus nonmotor processes during (visuo)motor control. *Human Brain Mapping*, **32**: 1311–1329.

Cameron, R. & Rogers, L. J. (1999). Hand preference of the common marmoset, problem solving and responses in a novel setting. *Journal of Comparative Psychology*, **113**: 149–157.

Cammarota, M., Bevilaqua, L. R., Rossato, J. I. *et al.* (2008). Parallel memory processing by the CA1 region of the dorsal hippocampus and the basolateral amygdala. *Proceedings of the National Academy of Sciences USA*, **105**: 10279–10284.

Canli, T., Desmond, J. E., Zhao, Z. et al. (2002). Sex differences in the neural basis of emotional memories. *Proceedings of the National Academy of Sciences USA*, **105**: 5532–5536.

Cantalupo, C. & Hopkins, W. D. (2010). The cerebellum and its contribution to complex tasks in higher primates: A comparative perspective. *Cortex*, **46**: 821–830.

Cantalupo, C., Bisazza, A. & Vallortigara, G. (1995). Lateralization of predator-evasion response in a teleost fish. *Neuropsychologia*, **33**: 1637–1646.

Cantalupo, C., Freeman, H., Rodes, W. & Hopkins, W. D. (2008). Handedness for tool use correlates with cerebellar asymmetries in chimpanzees (*Pan troglodytes*). *Behavioral Neuroscience*, **122**: 191–198.

Cantalupo, C., Pilcher, D. L. & Hopkins, W. D. (2003). Are planum temporale and sylvian fissure asymmetries directly correlated? A MRI study in great apes. *Neuropsychologia*, **41**: 1975–1981.

Carlson, K. J., Stout, D., Jashashvili, T. et al. (2011). The endocasts of MH1, *Australopithecus sediba*. *Science*, **333**: 1402.

Caron, J.-B., Morris, S. C. & Shu, D. (2010). Tentaculate fossils from the Cambrian of Canada (British Columbia) and China (Yunnan) interpreted as primitive Deuterostomes. *PLoS One*, **5**, e9586–A201.

Carrasquillo, Y. & Gereau IV, R. W. (2008). Hemispheric lateralization of a molecular signal for pain modulation in the amygdala. *Molecular Pain*, **4**: 24.

Casey, M. B. (2005). Asymmetrical hatching behaviors: The development of postnatal motor laterality in three precocial bird species. *Developmental Psychobiology*, **47**: 123–135.

Casey, B. J., Getz, S. & Galvan, A. (2008). The adolescent brain. *Developmental Review*, **28**: 62–77.

Casperd, J. M. & Dunbar, R. I. M. (1996). Asymmetries in the visual processing of emotional cues during agonistic interactions in gelada baboons. *Behavioral Processses*, **37**: 57–65.

Castelli, F., Frith, C., Happé, F. et al. (2002). Autism, Asperger syndrome and brain mechanisms for the attribution of mental states to animated shapes. *Brain*, **125**: 1839–1849.

Chapman, J. P. & Chapman, L. J. (1987). Handedness of hypothetically psychosis-prone subjects. *Journal of Abnormal Psychology*, **96**: 89–93.

Charron, S. & Koechlin, E. (2010). Divided representation of concurrent goals in the human frontal lobes. *Science*, **328**: 360–363.

Cheng Chen-To, Chang Heng & Hsu Pang-Ta (1957). *Sung Dynasty Album Paintings*. Peking: Chinese Classic Art Publishing House.

Cherkin, A. (1969). Kinetics of memory consolidation: Role of amnesic treatment parameters. *Proceedings of the National Academy of Sciences USA*, **63**: 1094–1101.

Chi, R. P. & Snyder, A. W. (2011). Facilitate insight by non-invasive brain stimulation. *PLoS One*, **6**: e16655.

Chiandetti, C. & Vallortigara, G. (2009). Effects of embryonic light stimulation on the ability to discriminate left from right in the domestic chick. *Behavioural Brain Research*, **198**: 204–246.

Chiandetti, C., Regolin, L., Rogers, L. J. & Vallortigara, G. (2005). Effects of light stimulation in embryo on the use of position-specific and object-specific cues in binocular and monocular chicks (*Gallus gallus*). *Behavioural Brain Research*, **163**:

10–17. Erratum to this paper published in 2007, *Behavioural Brain Research*, **177**: 175.

Chura, L. R., Lombardo, M. V., Ashwin, E. *et al.* (2010). Organizational effects of fetal testosterone on human corpus callosum size and asymmetry. *Psychoneuroendocrinology*, **35**: 122–132.

Cipolla-Neto, J., Horn, G. & McCabe, B. J. (1982). Hemispheric asymmetry and imprinting: The effect of sequential lesions of the hyperstriatum ventrale. *Experimental Brain Research*, **48**: 22–27.

Clark, B. J. & Taube, J. S. (2009). Deficits in landmark navigation and path integration after lesions of the interpeduncular nucleus. *Behavioral Neuroscience*, **123**: 490–503.

Clark, B. J., Sarma, A. & Taube, J. S. (2009). Head direction cell instability in the anterior dorsal thalamus after lesions of the interpeduncular nucleus. *Journal of Neuroscience*, **29**: 493–507.

Clark, M. M., Robertson, R. K. & Galef, B. G. (1993). Intrauterine position effects on sexually dimorphic asymmetries of Mongolian gerbils: Testosterone, eye-opening, and paw preference. *Developmental Psychobiology*, **26**: 185–194.

Clayton, N. S. (1993). Lateralization and unilateral transfer of spatial memory in marsh tits. *Journal of Comparative Physiology A*, **171**: 799–806.

Clayton, N. S. & Krebs, J. R. (1994a). Hippocampal growth and attrition in birds affected by experience. *Proceedings of the National Academy of Sciences USA*, **91**: 7410–7414.

Clayton, N. S. & Krebs, J. R. (1994b). Memory for spatial and object-specific cues in food-storing and non-storing birds. *Journal of Comparative Psychology A*, **174**: 371–379.

Collins, D. W. & Kimura, D. (1997). A large sex difference on a two-dimensional mental rotation task. *Behavioral Neuroscience*, **111**: 845–849.

Collins, R. L. (1985). On the inheritance of the direction and the degree of asymmetry. In: S. D. Glick (ed.), *Cerebral Lateralization in Nonhuman Species*, New York: Academic Press, pp. 41–71.

Collins, R. L. (1991). Reimpressed selective breeding for lateralization of handedness in mice. *Brain Research*, **564**: 194–202.

Collins, R. L., Sargent, E. E. & Neumann, P. E. (1993). Genetic and behavioral tests of the McManus hypothesis relating response to selection for lateralization of handedness in mice to degree of heterozygosity. *Behavioral Genetics*, **23**: 413–421.

Colonnese, M. T., Kaminska, A., Minlebaev, M. *et al.* (2010). A conserved switch in sensory processing prepares developing neocortex for vision. *Neuron*, **67**: 480–498.

Concha, M. L. & Wilson, S. W. (2001). Asymmetry in the epithalamus of vertebrates. *Journal of Anatomy*, **199**: 63–84.

Concha, M. L., Signore, I. A. & Colombo, A. (2009). Mechanisms of directional asymmetry in the zebrafish epithalamus. *Seminars in Cell and Developmental Biology*, **20**: 498–509.

Cooper, R., Nudo, N., González, J. *et al.* (2011). Side-dominance of *Periplaneta americana* persists through antenna amputation. *Journal of Insect Behavior*, **24**: 175–185.

Corballis, M. C. (2002). *From Hand to Mouth: The Origins of Language*. Princeton, NJ: Princeton University Press.

Corballis, M. C. (2003). From mouth to hand: Gesture, speech, and the evolution of right-handedness. *Behavioral Brain Sciences*, **26**: 199–260.

Corballis, M. C. (2006). Cerebral asymmetry: A question of balance. *Cortex*, **42**: 117–118.

Corballis, M. C., Badzakova-Trajkov, G. & Häberling, I. S. (2012). Right hand, left brain: Genetic and evolutionary bases of cerebral asymmetries for language and manual action. *WIREs Cognitive Science*, **3**: 1–17.

Corballis, P. M., Fendrich, R., Shapley, R. M. & Gazzaniga, M. S. (1999). Illusory contour perception and amodal boundary completion: Evidence of a dissociation following callosotomy. *Journal of Cognitive Neuroscience*, **11**: 459–466.

Corbetta, M., Patel, G. & Shulman, G. L. (2008). The reorienting system of the human brain: From environment to theory of mind. *Neuron*, **58**: 306–324.

Cowell, P. E. & Denenberg, V. H. (2002). Development of laterality and the role of the corpus callosum in rodents and humans. In: L. J. Rogers and R. J. Andrew (eds.), *Comparative Vertebrate Lateralization*, Cambridge: Cambridge University Press, pp. 274–305.

Cowell, P. E., Waters, N. S. & Denenberg, V. H. (1997). The effects of early environment on the development of functional laterality in Morris maze performance. *Laterality*, **2**: 221–232.

Craig, A. D. (2005). Forebrain emotional asymmetry: A neuroanatomical basis? *Trends in Cognitive Sciences*, **912**: 566–571.

Craig, A. D. (2009). How do you feel – now? The anterior insula and human awareness. *Nature Reviews Neuroscience*, **10**: 59–70.

Crockford, C., Herbinger, I., Vigilant, L. *et al.* (2004). Wild chimpanzees produce group-specific calls: A case for vocal learning? *Ethology*, **110**: 221–243.

Crow, T. J. (1997). Schizophrenia as a failure of hemispheric dominance for language. *Trends in Neurosciences*, **20**: 339–343.

Da Costa, A. P., Leigh, A. E., Man, M. & Kendrick, K. M. (2004). Face pictures reduce behavioural, autonomic, endocrine and neural indices of stress and fear in sheep. *Proceedings of the Royal Society of London B: Biological Sciences*, **271**: 2077–2084.

da Guardia, S. N. F., Cohen, L. G., da Cunha Pinho, M. *et al.* (2010). Interhemispheric asymmetry of corticomotor excitability after chronic cerebellar infarcts. *Cerebellum*, **9**: 398–404.

Dadda, M. & Bisazza, A. (2006). Does brain asymmetry allow efficient performance of simultaneous tasks? *Animal Behaviour*, **72**: 523–529.

Dadda, M., Koolhaas, W. H. & Domenici, P. (2010). Behavioural asymmetry affects escape performance in a teleost fish. *Biology Letters*, **6**: 414–417.

Daisley, J. N., Rosa Salva, O., Regolin, L. & Vallortigara, G. (2011). Social cognition and learning mechanisms: Experimental evidence in domestic chicks. *Interaction Studies*, **12**: 208–232.

Daisley, J. N., Vallortigara, G. & Regolin, L. (2010). Logic in an asymmetrical (social) brain: Transitive inference in the young domestic chick. *Social Neuroscience*, **5**: 309–319.

Davatzikos, C. & Resnick, S. M. (1998). Sex differences in anatomic measures of inter-hemispheric connectivity: Correlations with cognition in women but not men. *Cerebral Cortex*, **8**: 635–664.

Davidoff, J., Goldstein, J. & Roberson, D. (2009). Nature versus nurture: The simple contrast. *Journal of Experimental Child Psychology*, **102**: 246–250.

Davidson, R. J. (1995). Cerebral asymmetry, emotion and affective style. In: R. J. Davidson & K. Hugdahl (eds.), *Brain Asymmetry*, Cambridge, MA: MIT Press, pp. 361–387.

Davison, A., Frend, H. T., Moray, C. *et al.* (2009). Mating behaviour in pond snails *Lymnaea stagnalis* is a maternally inherited, lateralized trait. *Biology Letters*, 5: 20–22.

de Boyer des Roches, A., Durier, V., Richard-Yris, M. A. *et al.* (2011). Differential outcomes of unilateral interferences at birth. *Biology Letters*, 7: 177–180.

de Gelder, B., Portois, G. & Weiskrantz, L. (2002). Fear recognition in the voice is modulated by unconsciously recognised affective pictures. *Proceedings of the National Academy of Sciences USA*, 99: 4121–4126.

de Gennaro, L., Bertini, M., Pauri, F. *et al.* (2004). Callosal effects of transcranial stimulation (TMS): The influences of gender and stimulus parameters. *Neuroscience Research*, 48: 129–137.

de Latude, M., Demange, M., Bec, P. *et al.* (2009). Visual laterality responses to different emotive stimuli by red-capped mangabeys *Cercocebus torquatus torquatus*. *Animal Cognition*, 12: 31–42.

De Santi, A., Sovrano, V. A., Bisazza, A. & Vallortigara, G. (2001). Mosquitofish display differential left- and right-eye use during mirror image scrutiny and predator inspection responses. *Animal Behaviour*, 61: 305–310.

Deckel, A. W. (1995). Laterality of aggressive responses in *Anolis. Journal of Experimental Zoology*, 272: 194–200.

Deckel, A. W. (1997). Effects of alcohol consumption on lateralized aggression in *Anolis carolinensis. Brain Research*, 756: 96–105.

Deckel, A. W. (1998). Hemispheric control of territorial aggression in *Anolis carolinensis*: Effects of mild stress. *Brain, Behavior and Evolution*, 51: 33–39.

Deckel, A. W. & Fugua, L. (1998). Effects of serotonergic drugs on lateralized aggression and aggressive displays in *Anolis carolinensis. Behavioural Brain Research*, 95: 227–232.

Deckel, A. W., Lillaney, R., Ronan, P. J. & Summers, C. H. (1998). Lateralized effects of ethanol on aggression and serotonergic systems in *Anolis carolinensis. Brain Research*, 807: 38–46.

Dehaene-Lambertz, G., Hertz-Pannier, L. & Dubois, J. (2006). Nature and nurture in language acquisition: Anatomical and functional brain-imaging studies in infants. *Trends in Neurosciences*, 29: 367–381.

Denenberg, V. H. (1981). Hemispheric laterality in animals and the effects of early experience. *Behavioral and Brain Sciences*, 4: 1–49.

Denenberg, V. H. (1984). Behavioural asymmetry. In: N. Geschwind & A. M. Galaburda (eds.), *Cerebral Dominance: The Biological Foundations*, Cambridge MA: Harvard University Press, pp. 114–133.

Denenberg, V. H. (2005). Behavioral asymmetry and reverse asymmetry in the chick and rat. *Behavioral Brain Sciences*, 28: 597.

Denenberg, V. H., Cowell, P. E., Fitch, R. H., Kertesz, A. & Kenner, G. H. (1991). Corpus callosum: Multiple parameter measurements in rodents and humans. *Physiology and Behavior*, 49: 433–437.

Deng, C. & Rogers, L. J. (2002a). Factors affecting the development of lateralisation in chicks. In: L. J. Rogers & R. J. Andrew (eds.), *Comparative Vertebrate Lateralization*, Cambridge: Cambridge University Press, pp. 206–246.

Deng, C. & Rogers, L. J. (2002b). Social recognition and approach in the chick: Lateralization and effect of visual experience. *Animal Behaviour*, 63: 697–706.

Denny, K. (2009). Handedness and depression: Evidence from a large population survey. *Laterality*, 14: 246–255.

Deruelle, C. & Fagot, J. (1997). Hemispheric lateralisation and global precedence effects in the processing of visual stimuli by humans and baboons (*Papio papio*). *Laterlity*, 2: 233–246.

DeYoung, C. G., Hirsh, J. B., Shane, M. S. *et al.* (2010). Testing predictions from personality neuroscience: Brain structure and the Big Five. *Psychological Science*, 21: 820–828.

Diamond, M. C. (1991). Hormonal effects on the development of cerebral lateralisation. *Psychoneuroendocrinology*, 16: 121–128.

Diba, K. & Buzsáki, G. (2007). Forward and reverse hippocampal place-cell sequences during ripples. *Nature Neuroscience*, 10: 1241–1242.

Diekamp, B., Prior, H. & Güntürkün, O. (1999). Functional lateralization, interhemispheric transfer and position bias in serial reversal learning in pigeons (*Columba livia*). *Animal Cognition*, 2: 187–196.

Diekamp, B., Regolin, L., Gunturkun, O. & Vallortigara, G. (2005). A left-sided visuospatial bias in birds. *Current Biology*, 15: R372–R373.

Dien, J. (2008). Looking both ways through time: The Janus model of lateralised cognition. *Brain and Cognition*, 67: 292–323.

Dimond, S. & Harries, R. (1984). Face touching in monkeys, apes and man: Evolutionary origins and cerebral asymmetry. *Neuropsychologia*, 22: 227–233.

Dimond, S. J., Farrington, L. & Johnson, P. (1976). Differing emotional response from right and left hemispheres. *Nature*, 261: 690–692.

Dong, X.-P., Donoghue, P. C. J. & Repetski, J. E. (2005). Basal tissue structure in the earliest euconodonts: Testing hypotheses of developmental plasticity in euconodont phylogeny. *Palaeontology*, 48: 411–421.

Donoghue, P. C. J., Forey, P. L. & Andridge, R. J. (2000). Conodont affinity and chordate phylogeny. *Biological Review*, 75: 191–251.

Doupe, A. J. & Kuhl, P. K. (2008). Birdsong and human speech: Common themes and mechanisms. In: H. P. Zeigler & P. Marler (eds.), *Neuroscience of Birdsong*, Cambridge: Cambridge University Press, pp. 5–31.

Downs, A. & Smith, T. (2004). Emotional understanding, cooperation and social behaviour in high-functioning children with autism. *Journal of Autism and Developmental Disorders*, 34: 625–635.

Drach, P. (1948). Embranchement des Céphalochordés. In: P.-P. Grassé (ed.), *Traité de Zoologie*, Paris: Masson et Cie, pp. 931–1037.

Drews, C. (1996). Contexts and patterns of injuries in free-ranging male baboons (*Papio cynocephalus*). *Behaviour*, 133: 443–474.

Duguid, W. P. (2010). The enigma of reversed asymmetry in lithodid crabs: Absence of evidence for heritability or induction of morphological handedness in *Lopholithodes foraminatus*. *Evolution and Development*, 12: 74–83.

Duistermars, B. J., Chow, D. M. & Frye, M. A. (2009). Flies require bilateral sensory input to track odour gradients in flight. *Current Biology* 19: 1301–1307.

Eaton, R. C. & Emberley, D. S. (1991). How stimulus direction determines the trajectory of the Mauthner-initiated escape response in a teleost fish. *Journal of Experimental Biology*, 161: 469–487.

Ehret, G. (1987). Left hemisphere advantage in the mouse brain for recognising ultra-sonic communication calls. *Nature*, **325**: 249–251.

Ehrlichman, H. (1986). Hemispheric asymmetry and positive-negative affect. In: D. Ottoson (ed.), *Duality and Unity of the Brain*, Dordrecht: The Netherlands, pp. 194–206.

Engbretson, G. A., Reiner, A. & Brecha, N. (1981). Habenular asymmetry and the central connections of the parietal eye of the lizard. *Journal of Comparative Neurology*, **198**: 155–165.

Enggist-Dueblin, P. & Pfister, U. (2002). Cultural transmission of vocalisations in ravens *Corvus corax*. *Animal Behaviour*, **64**: 831–841.

Erwin, C. W. & Linnoila, M. (1981). Effect of ethyl alcohol on visual evoked potentials. *Alcoholism: Clinical and Experimental Research*, **5**: 49–55.

Esslinger, C., Kirsch, P. & Haddad, L. *et al.* (2011). Cognitive state and connectivity effects of the genome-wide significant psychosis variant in ZNF804A. *NeuroImage*, **54**: 2514–2523.

Everhart, D. E., Suchard, J. L., Quatrin, T. *et al.* (2001). Sex-related difference in event-related potentials, face recognition and facial affect processing in prepubertal children. *Neuropsychology*, **15**: 329–341.

Fagot, J. & Vauclair, J. (1991). Manual laterality in nonhuman primates: A distinction between handedness and manual specialization. *Psychological Bulletin*, **109**: 76–89.

Fagot, J., Lacreuse, A. & Vauclair, J. (1997). Role of sensory and post-sensory factors on hemispheric asymmetries in tactual perception. In: S. Christman (ed.), *Cerebral Asymmetries in Sensory and Perceptual Processing*, New York: Elsevier, pp. 469–494.

Fan, J., Gu, X., Guise, K. G. *et al.* (2011). Involvement of the anterior cingulate and frontoinsular cortices in rapid processing of salient emotional information. *NeuroImage*, **54**: 2539–2546.

Fan, L., Tang, Y., Sun, B. *et al.* (2010). Sexual dimorphism and asymmetry in human cerebellum: An MRI-based morphometric study. *Brain Research*, **1353**: 60–73.

Faurie, C. & Raymond, M. (2004). Handedness, homicide and negative frequency-dependent selection. *Proceedings of the Royal Society of London B*, **272**: 25–28.

Faurie, C. & Raymond, M. (2005). Handedness frequency over more than 10,000 years. *Proceedings of the Royal Society of London B*, **271**: S43–S45.

Ferbinteanu, J. & Shapiro, M. L. (2003). Prospective and retrospective memory coding in the hippocampus. *Neuron*, **40**: 1227–1239.

Fernandez-Carriba, S., Loches, A. & Hopkins, W. D. (2002). Asymmetry of facial expression of emotions by chimpanzees. *Neuropsychologia*, **40**: 1523–1533.

Ferrari, P. F., Paukner, A., Ruggiero, A. *et al.* (2009). Inter-individual differences in neonatal imitation and the development of action chains in rhesus macaques. *Child Development*, **80**: 1057–1068.

Ficken, M. S. (1977). Avian play. *The Auk*, **94**: 573–582.

Finch, G. (1941). Chimpanzee handedness. *Science*, **94**: 117–118.

Fitch, R. H., Berrebi, A. S., Cowell, P. E., Schrott, L. M. & Denenberg, V. H. (1990). Corpus callosum: Effects of neonatal hormones on sexual dimorphism in the rat. *Brain Research*, **515**: 111–116.

Fitch, R. H., Cowell, P. E., Schrott, L. M. & Denenberg, V. H. (1991). Corpus callosum: Demasculinization via perinatal anti-androgen. *International Journal of Developmental Neuroscience*, 1: 35–38.

Foa, A., Basaglia, G., Carnacina, M. *et al.* (2009). Orientation of lizards in a Morris water-maze: Roles of the sun compass and the parietal eye. *Journal of Experimental Biology*, 212: 1918–2924.

Folta, K., Diekamp, B. & Güntürkün, O. (2004). Asymmetrical modes of visual bottom-up and top-down integration in the thalamic nucleus rotundus of pigeons. *Journal of Neuroscience*, 24: 9475–9485.

Forrester, G. S., Quaresmini, C., Leavens, D. A., Spiezio, C. & Vallortigara, G. (2012). Target animacy influences chimpanzee handedness. *Animal Cognition*, advance online publication at http://www.ncbi.nlm.nih.gov/pubmed/22829099.

Forrester, G. S., Quaresmini, C., Leavens, D. A. & Vallortigara, G. (2011). Target animacy influences gorilla handedness. *Animal Cognition*, 14: 903–907.

Foster, W. A. & Treherne, J. E. (1981). Evidence for the dilution effect in the selfish herd from fish predation of a marine insect. *Nature*, 293: 508–510.

Foundas, A. L., Leonard, C. M. & Hanna-Pladdy, B. (2002). Variability in the anatomy of the planum temporale and posterior ascending ramus: Do right and left handers differ? *Brain and Language*, 83: 403–424.

Fox, M. D., Corbetta, M., Snyder, A. Z. *et al.* (2006). Spontaneous neural activity distinguishes human dorsal and ventral attention systems. *Proceedings of the National Academy of Sciences USA*, 103: 10046–10051.

Franklin, A. (2009). Pre-linguistic categorical perception of colour cannot be explained by colour preference: Response to Roberson and Hanley. *Trends in Cognitive Sciences*, 13: 501–502.

Franklin, A., Drivonikou, G. V., Bevis, L. *et al.* (2008a). Categorical perception of colour is lateralised to the right hemisphere in infants, but to the left hemisphere in adults. *Proceedings of the National Academy of Sciences USA*, 105: 3221–3225.

Franklin, A., Drivonikou, G. V., Clifford, A. *et al.* (2008b). Lateralisation of categorical perception colour changes with colour term acquisition. *Proceedings of the National Academy of Sciences USA*, 105: 18221–18225.

Frasnelli, E., Anfora, G., Trona, F., Tessarolo, F. & Vallortigara, G. (2010). Morpho-functional asymmetry of the olfactory receptors of the honeybee (*Apis mellifera*). *Behavioural Brain Research*, 209: 221–225.

Frasnelli, E., Iakovlev, I. & Reznikova, Z. (2012). Asymmetry in antennal contacts during trophallaxis in ants. *Behavioural Brain Research*, 32: 7–12.

Frasnelli, E., Vallortigara, G. & Rogers, L. J. (2011). Origins of brain asymmetry: Lateralization of odour memory recall in primitive Australian stingless bees. *Behavioural Brain Research*, 224: 121–127.

Frasnelli, E., Vallortigara, G., Rogers, L. J. (2012). Left–right asymmetries of behaviour and nervous system in invertebrates. *Neuroscience and Biobehavioral Reviews*, 36: 1273–1291.

Freake, M. J. (1999). Evidence for orientation using the e-vector of polarised light in the sleepy lizard *Tiliqua rugosa*. *Journal of Experimental Biology*, 202: 1159–1166.

Freake, M. J. (2001). Homing behaviour in the sleepy lizard *Tiliqua rugosa*: The role of visual cues and the parietal eye. *Behavioral Ecology and Sociobiology*, 50: 563–569.

Fredes, F., Tapia, S., Letelier, J. C. *et al.* (2010). Topographical arrangement of the rotundo-entopallial projection in the pigeon (*Columba livia*). *Journal of Comparative Neurology*, **518**: 4342–4361.

Freire, R. & Rogers, L. J. (2005). Experience-induced modulation of the use of spatial information in the domestic chick. *Animal Behaviour*, **69**: 1093–1100.

Freire, R. & Rogers, L. J. (2007). Experience during a period of right hemispheric dominance alters attention to spatial; information in the domestic chick. *Animal Behaviour*, **74**: 413–418.

Freire, R., Cheng, H.-W. & Nicol, C. J. (2004). Development of spatial memory in occlusion-experienced domestic chicks. *Animal Behaviour*, **67**: 141–150.

Freire, R., van Dort, S. & Rogers, L. J. (2006). Pre- and post- hatching effects of corticosterone treatment on behavior of the domestic chick. *Hormones and Behavior*, **49**: 157–165.

Freund, N., Güntürkün, O. & Manns, M. (2008). A morphological study of the nucleus subpretectalis of the pigeon. *Brain Research Bulletin*, **75**: 491–493.

Friedrich, A. & Teyke, T. (1998). Identification of stimuli and input pathways mediating food-attraction conditioning in the snail *Helix*. *Journal of Comparative Physiology A*, **183**: 247–254.

Frith, E. L. & Frith, U. (2003). Understanding autism: Insights from mind and brain. *Philosophical Transactions of the Royal Society of London B*, **358**: 281–289.

Fu, C. H., Vythelingum, G. N., Brammer, M. J. *et al.* (2006). An fMRI study of verbal self-monitoring: Neural correlates of auditory verbal feedback. *Cerebral Cortex*, **16**: 969–977.

Gainotti, G. (1972). Emotional behaviour and hemispheric side of the lesion. *Cortex*, **8**: 41–55.

Gainotti, G. (1989). Disorders of emotions and affect in patients with unilateral brain damage. In: F. Boller & J. Grafman (eds.), *Handbook of Neuropsychology*, Vol. 3, Amsterdam: Elsevier, pp. 161–179.

Gallate, J., Wong, C., Ellwood, S., Chi, R. & Snyder, A. (2011). Noninvasive brain stimulation reduces prejudice scores on an implicit association test. *Neuropsychology*, **25**: 185–192.

Gardner, R. A., Vancante, T. E. & Gardner, B. T. (1992). Categorical replies to categorical questions by cross-fostered chimpanzees. *American Journal of Psychology*, **105**: 27–57.

Gazzaniga, M. (1967). The split-brain in man. *Scientific American*, **217**: 24.

Gazzaniga, M. S. (2000). Cerebral specialisation and interhemispheric communication. *Brain*, **123**: 1293–1326.

Geissler, D. B. & Ehret, G. (2004). Auditory perception vs. recognition: Representation of complex communication in the mouse auditory fields. *European Journal of Neuroscience*, **19**: 1027–1040.

Geng, J. J. & Mangun, C. R. (2011). Right temporoparietal junction activation by a salient contextual cue facilitates target discrimination. *NeuroImage*, **54**: 594–601.

Gentilucci, M., Benuzzi, F., Gangitano, M. & Grimaldi, S. (2001). Grasp with hand and mouth: A kinematic study on healthy subjects. *Journal of Neurophysiology*, **86**: 1685–1699.

George, I., Vernier, B., Richard, J.-P., Hausbeger, M. & Cousillas, H. (2004). Hemispheric specialization in the primary auditory area of awake and anesthetized starlings (*Sturnus vulgaris*). *Behavioral Neuroscience*, **118**: 597–610.

Geschwind, N. & Galaburda, A. M. (1987). *Cerebral Lateralization: Biological Mechanisms, Associations, and Pathology*. Cambridge, MA: MIT Press.

Ghirlanda, S. & Vallortigara, G. (2004). The evolution of brain lateralization: A game theoretical analysis of population structure. *Proceedings of the Royal Society of London B*, **271**: 853–857.

Ghirlanda, S., Frasnelli, E. & Vallortigara, G. (2009). Intraspecific competition and coordination in the evolution of lateralization. *Philosophical Transactions of the Royal Society of London B*, **364**: 861–866.

Gianotti, L. R. R., Knoch, D., Faba, P. L. *et al.* (2009). Tonic activity level in right prefrontal predicts individuals' risk taking. *Psychological Science*, **20**: 33–38.

Gibbs, M. E., Andrew, R. J. & Ng, K. T. (2003). Hemispheric lateralisation of memory stages for discriminated avoidance learning in the chick. *Behavioural Brain Research*, **139**: 157–165.

Gilbert, A. L., Regier, T., Kay, P. *et al.* (2006). Whorf hypothesis is supported in the right visual field but not the left. *Proceedings of the National Academy of Sciences USA*, **103**: 489–494.

Gilby, I. C. (2006). Meat sharing amongst the Gombe chimpanzees: Harassment and reciprocal exchange. *Animal Behaviour*, **71**: 953–963.

Giljov, A., Karenina, K. & Malashichev, Y. (2012). Limb preferences in a marsupial, *Macropus rufogriseus*: Evidence for postural effect. *Animal Behaviour*, **83**, 525–534.

Giljov, A. N., Karenina, K. A. & Malashichev, Y. B. (2009). An eye for a worm: Lateralisation of feeding behaviour in aquatic anamniotes. *Laterality*, **14**: 273–286.

Goldstein, K. (1939/1963). *The Organism: A Holistic Approach to Biology*. New York: The American Book Co.

Gomez, M. del P., Angucyra, J. M. & Nasi, E. (2009). Light-transduction in melanopsin-expressing photoreceptors of Amphioxus. *Proceedings of the National Academy of Sciences USA*, **106**: 9081–9086.

Gordon, D. J. & Rogers, L. J. (2010). Differences in social and vocal behavior between left- and right-handed common marmosets. *Journal of Comparative Psychology*, **124**: 402–411.

Gorrie, C. A., Waite, P. M. & Rogers, L. J. (2008). Correlations between hand preference and cortical thickness in the secondary somatosensory (SII) cortex of the common marmoset, *Callithrix jacchus*. *Behavioral Neuroscience*, **122**: 1343–1351.

Goto, K., Kurashima, R., Gokan, H. *et al.* (2010). Left–right asymmetry defect in the hippocampal circuitry impairs spatial dexterity and working memory in *iv* mice. *PLoS One*, **5**(11): e15468. doi: 10.1371/journal.pone.0015468.

Gottlieb, G. (2002). On the epigenetic evolution of species-specific perception: The developmental manifold concept. *Cognitive Development*, **17**: 1287–1300.

Govind, C. K. (1992). Claw asymmetry in lobsters: Case study in developmental neuro-ethology. *Journal of Neuroethology*, **23**: 1423–1445.

Grace, J. K. & Craig, D. P. (2008). The development of lateralization of prey delivery in a bill load holding bird. *Animal Behaviour*, **75**: 2005–2011.

Grande, C. & Patel, N. H. (2009). Nodal signalling is involved in left–right asymmetry in snails. *Nature*, **457**: 1007–1011.

Greicius, M. D., Krasnow, B., Reiss, A. L. *et al.* (2003). Functional connectivity in the resting brain: A network analysis of the default network hypothesis. *Proceedings of the National Academy of Sciences USA*, **100**: 253–258.

Greicius, M. D., Srivastava, G., Reiss, A. L. et al. (2004). Default-mode network activity distinguishes Alzheimer's disease from healthy aging: Evidence from functional MRI. *Proceedings of the National Academy of Sciences USA*, **101**: 4637–4642.

Grimm, S., Beck, J., Schuepbach, D. et al. (2008). Imbalance between left and right dorsolateral prefrontal cortex in major depression is linked to negative emotional judgment: An fMRI study in severe major depressive disorder. *Biological Psychiatry*, **63**: 369–376.

Groothuis, T. G. & Schwabl, H. (2002). Determinants of within- and among-clutch variation in levels of maternal hormones in Black-Headed Gull eggs. *Functional Ecology*, **16**: 281–289.

Guenther, F. H. (2006). Cortical interactions underlying the production of speech sounds. *Journal of Communicative Disorders*, **39**: 350 –365.

Guglielmotti, V. & Cristino, L. (2006). The interplay between the pineal complex and the habenular nuclei in lower vertebrates in the context of the evolution of cerebral asymmetry. *Brain Research Bulletin*, **69**: 475–488.

Guiard, Y., Diaz, G. & Beaubaton, D. (1983). Left hand advantage in right handers for spatial constant error: Preliminary evidence in a unimanual ballistic aimed movement. *Neuropsychologia*, **21**: 111–115.

Guioli, S. & Lovell-Badge, R. (2007). PITX2 controls asymmetric gonadal development in both sexes of the chick and can rescue the degeneration of the right ovary. *Development*, **134**: 4199–4208.

Güntürkün, O. (1993). The ontogeny of visual lateralization in pigeons. *German Journal of Psychology*, **17**: 276–287.

Güntürkün, O. (2002). Ontogeny of visual asymmetry in pigeons. In: L. J. Rogers & R. J. Andrew (eds.), *Comparative Vertebrate Lateralization*, Cambridge: Cambridge University Press, pp. 247–273.

Güntürkün, O. (2003). Adult persistence of head-turning asymmetry. *Nature*, **421**: 711.

Güntürkün, O. & Kesh, S. (1987). Visual lateralization during feeding in pigeons. *Behavioural Neuroscience*, **101**: 433–435.

Güntürkün, O., Diekamp, B., Manns, M. et al. (2000). Asymmetry pays: Visual lateralization improves discrimination success in pigeons. *Current Biology*, **10**: 1079–1081.

Guo, K., Meints, K., Hall, C., Hall, S. & Mills, D. (2009). Left gaze bias in humans, rhesus monkeys and domestic dogs. *Animal Cognition*, **12**: 409–418.

Gutiérrez-Ibáñez, C., Reddon, A. R., Kreuzer, M. B., Wylie, D. R. & Hurd, P. L. (2011). Variation in asymmetry of the habenular nucleus correlates with behavioural asymmetry in a cichlid fish. *Behavioural Brain Research*, **221**: 189–196.

Gutnick, T., Byrne, R. A., Hochner, B. et al. (2011). *Octopus vulgaris* uses visual information to determine location of its arms. *Current Biology*, **21**: 460–462.

Güven, M., Elalmis, D. D., Binokay, S. & Tan, U. (2003). Population-level right-paw preference in rats assessed by a new computerized food-reaching test. *International Journal of Neuroscience*, **113**: 1675–1689.

Haakonsson, J. E. & Semple, S. (2009). Lateralisation of trunk movements in captive Asian elephants (*Elephas maximus*). *Laterality*,**14**: 413–422.

Haase, A., Rigosi, E., Trona, F. et al. (2011a). In-vivo two-photon imaging of the honeybee antennal lobe. *Biomedical Optics Express*, **2**: 131–138.

Haase, A., Rigosi, E., Trona, F. *et al.* (2011b). A multimodal approach for tracing lateralisation along the olfactory pathway in the honeybee through electrophysiological recordings, morpho-functional imaging, and behavioural studies. *European Biophysics Journal with Biophysics Letters*, 40: 1247–1258.

Habas, C., Kamdar, N., Nguyen, D. *et al.* (2009). Distinct cerebellar contributions to intrinsic connectivity networks. *Journal of Neuroscience*, 29: 8586–8594.

Häberling, I. S., Badzakova-Trajko, G. & Corballis, M. C. (2011). Callosal tracts and patterns of hemispheric dominance: A combined fMRI and DTI study. *NeuroImage*, 54: 779–786.

Hall, J. (2008). *The Sinister Side: How Left–Right Symbolism Shapted Western Art.* Oxford: Oxford University Press.

Halpern, M. E., Güntürkün, O., Hopkins, W. D. & Rogers, L. J. (2005). Lateralization of the vertebrate brain: Taking the side of model systems. *Journal of Neuroscience*, 9: 10351–10357.

Halpern, M. E., Liang, J. O. & Gamse, J. T. (2003). Leaning to the left: Laterality in the zebrafish forebrain. *Trends in Neurosciences*, 26: 308–313.

Hamilton, C. R. (1988). Hemispheric specialization in monkeys. In: C. Trevarthen (ed.), *Brain Circuits and Functions of the Mind*, Cambridge: Cambridge University Press, pp. 181–195.

Hamilton, C. R. & Vermeire, B. A. (1988). Complementary hemispheric specialization in monkeys. *Science*, 242: 1691–1694.

Hampson, E. (1990). Estrogen-related variation in human spatial and articulatory motor skills. *Psychoneuroendocrinology*, 15: 97–111.

Hardyck, C., Goldman, R. & Petrinovich. L. (1975). Handedness and sex, race, and age. *Human Biology*, 47: 369–375.

Harmon-Jones, E., Gable, P. A. & Peterson, C. K. (2010). The role of asymmetric frontal cortical activity in emotion-related phenomena: A review and update. *Biological Psychiatry*, 84: 451–462.

Harris, L. J. (1989). Footedness in parrots: Three centuries of research, theory, and mere surmise. *Canadian Journal of Psychology*, 43: 369–396.

Harvey, C. (2011). Humanity's first word? Duh! *New Scientist*, 26 November, p. 10.

Hazlerigg, D. & Loudon, A. (2008). New insights into ancient seasonal times. *Current Biology*, 18: R795–R804.

Hecht, D. (2011). An inter-hemispheric imbalance in the psychopath's brain. *Personality and Individual Differences*, 51: 3–10.

Hellige, J. B. (1993a). *Hemispheric Asymmetry: What's Right and What's Left.* Cambridge, MA: Harvard University Press.

Hellige, J. B. (1993b). Unity of thought and action: Varieties of interaction between the left and right hemispheres. *Current Directions in Psychological Sciences*, 2: 21–25.

Herlitz, A., Nilsson, L.-G. & Backman, L. (1997). Gender differences in episodic memory. *Memory and Cognition*, 25: 801–811.

Heuts, B. A. (1999). Lateralization of trunk muscle volume, and lateralization of swimming turns of fish responding to external stimuli. *Behavioral Processes*, 47: 113–124.

Heuts, B. A. & Brunt, T. (2005). Behavioural left–right asymmetry extends to arthropods. *Behavioural Brain Science*, 28: 601–602.

Heuts, B. A., Cornelissen, P. & Lambrechts, D. Y. M. (2003). Different attack modes of Formica species in interspecific one-on-one combats with other ants (*Hymenoptera: Formicidae*). *Annals Zoology (Wars)*, **53**: 205–216.

Hewes, G. W. (1976). Current status of gestural theory of language origin. *Annals of the New York Academy of Science*, **280**: 482–504.

Hews, D. K. & Worthington, R. A. (2001). Fighting from the right side of the brain: Left visual field preference during aggression in free-ranging male tree lizards (*Urosaurus ornatus*). *Brain Behavior and Evolution*, **58**: 356–361.

Hews, D. K., Castellano, M. & Hara, E. (2004). Aggression in females is also lateralized: Left-eye bias during aggressive courtship rejection in lizards. *Animal Behaviour*, **68**: 1201–1207.

Hickok, G., Costanzo, M., Capasso, R. *et al.* (2011). The role of Broca's area in speech perception: Evidence from aphasia revisited. *Brain and Language*, **119**: 214–220.

Higuchia, S., Chaminade, T., Imanizu, H. *et al.* (2009). Shared neural correlates for language and tool use in Broca's area. *NeuroReport*, **20**: 1376–1381.

Hikosaka, O. (2010). The habenula: From stress evasion to value-based decision-making. *Nature Neuroscience*, **11**: 503–513.

Hill, A., Howard, C. V., Strahle, U. & Cossins, A. (2003). Neurodevelopmental defects in zebrafish (*Danio rerio*) at environmentally relevant dioxin (TCDD) concentrations. *Toxicology Science*, **76**: 392–399.

Hill, K. R., Walker, R. S., Božičevíc, M. *et al.*(2011). Co-residence patterns in hunter-gatherer societies show unique human social structure. *Science*, **331**: 1286–1289.

Hirnstein, M., Leask, S., Rose, J. & Hausmann, M. (2010). Disentangling the relationship between hemispheric asymmetry and cognitive performance. *Brain and Cognition*, **73**: 119–127.

Hobert, O., Johnston, R. J. Jr & Chang, S. (2002). Left-right asymmetry in the nervous system: The *Caenorhabditis elegans* model. *Nature Reviews Neuroscience*, **3**: 629–640.

Hochner, B., Shomrat, T. & Fiorito, G. (2006). The octopus: A model for the comparative analysis of the evolution of learning and memory. *Biological Bulletin*, **210**: 308–317.

Hodos, W. & Campbell, C. B. G. (1969). Scala Naturae: Why there is no theory in comparative psychology. *Psychological Review*, **76**: 337–350.

Hoffman, A. M., Robakiewicz, P. E., Tuttle, E. M. & Rogers, L. J. (2006). Behavioural lateralization in the Australian magpie (*Gymnorhina tibicen*). *Laterality*, **11**: 110–121.

Holdstock, J. S., Crane, J., Bachorowski, J. A. *et al.* (2010). Equivalent activation of the hippocampus by face–face and face–laugh paired associate learning and recognition. *Neuropsychologia*, **48**: 3757–3771.

Holland, R., Leff, A. P., Josephs, O. *et al.* (2011). Speech facilitation by left inferior frontal cortex stimulation. *Current Biology*, **21**: 1403–1407.

Hook, M. A. & Rogers, L. J. (2000). Development of hand preferences in marmosets (*Callithrix jacchus*) and effects of ageing. *Journal of Comparative Psychology*, **114**: 263–271.

Hook, M. A. & Rogers, L. J. (2008). Visuospatial reaching preferences of common marmosets: An assessment of individual biases across a variety of tasks. *Journal of Comparative Psychology*, **122**: 41–51.

Hopkins, W. D. (1995). Hand preferences for a coordinated bimanual task in 110 chimpanzees (*Pan troglodytes*): Cross-sectional analysis. *Journal of Comparative Psychology*, **109**: 291–297.

Hopkins, W. D. (1997). Hemispheric specialisation for local and global processing of hierarchical visual stimuli in chimpanzees (Pan troglodytes). Neuropsychologia, 35: 343–348.

Hopkins, W. D. (2006). Comparative and familial analysis of handedness in great apes. Psychological Bulletin, 132: 538–559.

Hopkins, W. D. (ed.) (2007). Evolution of Hemispheric Specialization in Primates. Oxford: Academic Press.

Hopkins, W. D. & Bennett, A. J. (1994). Handedness and approach-avoidance behaviour in chimpanzees (Pan troglodytes). Journal of Experimental Psychology, 20: 413–418.

Hopkins, W. D. & Cantalupo, C. (2004). Handedness in chimpanzees (Pan troglodytes) is associated with asymmetries of the primary motor cortex but not with homologous language areas. Behavioral Neuroscience, 118: 1176–1183.

Hopkins, W. D. & Nir, T. M. (2010). Planum temporale surface area and grey matter asymmetries in chimpanzees (Pan troglodytes): The effect of handedness and comparison with findings in humans. Behavioural Brain Research, 208: 436–443.

Hopkins, W. D., Phillips, K. A., Bania, A., Calcutt et al. (2011). Hand preferences for coordinated bimanual actions in 777 great apes: Implications for the evolution of handedness in Hominins. Journal of Human Evolution, 60: 605–611.

Hopkins, W. D., Russell, J. L. & Cantalupo, C. (2007). Neuroanatomical correlates of handedness for tool use in chimpanzees (Pan troglodytes): Implication for theories on the evolution of language. Psychological Sciences, 18: 971–977.

Hopkins, W. D., Russell, J. L., Freeman, H. et al. (2006). Lateralized scratching in chimpanzees: Evidence of a functional asymmetry during arousal. Emotion, 6: 553–559.

Hopkins, W. D., Russell. J. L., Schaeffer, J. A. et al. (2009). Handedness for tool use in captive chimpanzees (Pan troglodytes): Sex differences, performance, heritability and comparison to the wild. Behaviour, 146: 1463–1483.

Hopkins, W. D., Wesley, M. J., Izard, M. K., Hook, M. & Schapiro, S. J. (2004). Chimpanzees are predominantly right-handed: Replication in three colonies of apes. Behavioral Neuroscience, 118: 659–663.

Hopp, S. L., Jablonski, P. & Brown, J. L. (2001). Recognition of group membership by voice in Mexican jays, Aphelocoma ultramarina. Animal Behaviour, 62: 297–303.

Hori, M. (1993). Frequency-dependent natural selection in the handedness of scale-eating cichlid fish. Science, 260: 216–219.

Horn, G. (1985). Memory, Imprinting and the Brain. Oxford: Clarendon Press.

Horn, G. (1991). Cerebral Function and Behaviour Investigated through a Study of Filial Imprinting. Cambridge: Cambridge University Press.

Horn, G. (2004). Pathways of the past; the imprint of memory. Nature Reviews Neuroscience, 5: 108–120.

Horn, G., Rose, S. P. R. & Bateson, P. P. G. (1973). Experience and plasticity in the central nervous system. Science, 181: 506–514.

Horowitz, A. (2009). Attention to attention in domestic dog (Canis familiaris). Animal Cognition, 12: 107–118.

Hostetter, A. B., Cantero, M. & Hopkins, W. D. (2001). Differential use of vocal and gestural communication by chimpanzees (Pan troglodytes) in response to attentional status of a human (Homo sapiens). Journal of Comparative Psychology, 115: 337–343.

Hourcade, B., Perisse, E., Devaud. J.-M. *et al.* (2009). Long-term memory shapes the primary olfactory centre of an insect brain. *Learning and Memory*, **16**: 607–615.

Howard, R. J., Ffytche, D. H., Barnes, J. *et al.* (1998). The functional anatomy of imaging and perceiving colour. *NeuroReport*, **9**: 1019–1023.

Hugdahl, K. (1995). Classical conditioning and implicit learning: The right hemisphere hypothesis. In: R. J Davidson & K. Hugdahl (eds.), *Brain Asymmetry*, Cambridge, MA: MIT Press, pp. 235–267.

Hui-Di, Y., Wang, Q., Wang, Z. *et al.* (2011). Food hoarding and associated neuronal activation in brain reward circuitry in Mongolian gerbils. *Physiology and Behavior*, **104**: 429–436.

Humphrey, N. (1998). Left-footedness in peacocks: An emperor's tale. *Laterality*, **3**: 289–289.

Huster, R. J., Westerhausen, R. & Herrmann, C. S. (2011). Sex differences in cognitive control are associated with midcingulate and callosal morphology. *Brain Structure and Function*, **215**: 225–235.

Iacoboni, M., Molnar-Szakaes, I., Gallese, V. *et al.* (2005). Grasping the intentions of others with one's own mirror neuron system. *PLoS Biology*, **3**: e79.

Ingle, D. J. & Hoff, K. vS. (1990). Visually evoked evasive behaviour in frogs. *BioScience*, **40**: 284–291.

Iturria-Medina, Y., Péréz Fernández, A., Morris, D. M. *et al.* (2011). Brain hemispheric structural efficiency and interconnectivity rightward asymmetry in human and non-human primates. *Cerebral Cortex*, **21**: 56–67.

Izquierdo, I., Bevilaqua, L. R. M., Rossato, J. I. *et al.* (2006). Different molecular cascades in different sites of the brain control memory consolidation. *Trends in Neuroscience*, **29**: 496–505.

Jacobs, L. F. & Spencer, W. D. (1994). Natural space-use patterns and hippocampal size in kangaroo rats. *Brain, Behavior and Evolution*, **44**: 125–132.

James, T. W. & Kimura, D. (1997). Sex differences in remembering the locations of objects in an array: Location-shift versus location-exchanges. *Evolution and Human Behavior*, **18**: 155–163.

Jamieson, D. & Roberts, A. (1999). A possible pathway connecting the photosensitive pineal eye to the swimming generator in young *Xenopus laevis* tadpoles. *Brain Behaviour and Evolution*, **54**: 323–337.

Jefferies, R. P. S. & Lewis, D. N. (1978). The English Silurian fossil *Placocystites forbesianus* and the ancestry of the vertebrates. *Philosophical Transactions of the Royal Society of London B*, **282**: 205–323.

Johanson, R. S., Theorin, A., Westling, G. *et al.* (2006). How a lateralised brain supports symmetrical bimanual tasks. *PLoS Biol.*, **4**: 1462–1466.

Johnson, K. M., Boonstra, R. & Wojtowicz, J. M. (2010). Hippocampal neurogenesis in food-storing red squirrels: The impact of age and spatial behavior. *Genes, Brain and Behavior*, **9**: 583–591.

Johnston, A. N. B. & Rose, S. P. R. (2002). Memory and lateralised recall. In: L. J. Rogers & R. J. Andrew (eds.), *Comparative Vertebrate Lateralisation*, Cambridge: Cambridge University Press, pp. 533–581.

Jozet-Alves, C., Romagny, S., Bellanger, C. & Dickel, L. (2012). Cerebral correlates of visual lateralization in *Sepia*. *Behavioural Brain Research*, **234**: 20–25.

Kanwisher, N., Chunn, M. M., McDermott, J. & Ledden, P. J. (1996). Functional imaging of human visual recognition. *Cognitive Brain Research*, **5**: 55–67.

Kaplan, G. (2000). Song structure and function of mimicry in the Australian magpie (*Gymnorhina tibicen*) compared to the lyrebird (*Menura ssp.*). *International Journal of Comparative Psychology*, **12**: 219–241.

Kaplan, G. (2008). The Australian magpie (*Gymnorhina tibicen*): An alternative model for the study of songbird neurobiology. In: P. Zeigler & P. Marler (eds.), *The Neuroscience of Birdsong*, Cambridge: Cambridge University Press, pp. 153–170.

Kaplan, G. & Rogers, L. J. (2002). Patterns of gazing in orang-utans (*Pongo pygmaeus*). *International Journal of Primatology*, **23**: 501–526.

Kaplan, G. & Rogers, L. J. (2003). *Gene Worship: Moving Beyond the Nature/Nurture Debate over Genes, Brain, and Gender*. New York: OtherPress.

Kaplan, G., Pines, M. K. & Rogers, L. J. (2012). Stress and stress reduction in common marmosets. *Applied Animal Behaviour Science*, **137**: 175–182.

Kappers, A., Huber, G. C. & Crosby, E. A. (1936). *The Comparative Anatomy of the Nervous System of Vertebrates, Including Man*. New York: Hafner Publishing Company.

Karenina, K., Giljov, A., Baranov, V. et al. (2010). Visual laterality of calf–mother interactions in wild whales. *PLoS One*, **5**: e13787.

Kawakami, R., Dobe, A., Shibemoto, R. et al. (2008). Right isomerism of the brain in *inversus viscerum* mutant mice. *PLoS One*, **3**(4): e1945.

Kawakami, R., Shinohara, Y., Kato, Y. et al. (2003). Asymmetric allocation of NMDA receptor ε2 subunits in hippocampal circuitry. *Science*, **300**: 990–994.

Keenan, J. P., Nelson, A., O'Connor, M. & Pascual-Leone, A. (2001). Self recognition and the right hemisphere. *Nature*, **409**: 305.

Kells, A. R. & Goulson, D. (2001). Evidence for handedness in bumblebees. *Journal of Insect Behaviour*, **14**: 47–55.

Kendrick, K. M. (2006). Brain asymmetries for face recognition and emotion control in sheep. *Cortex*, **42**: 96–98.

Kendrick, K. M. & Baldwin, B. A. (1987). Cells in the temporal cortex of sheep can respond preferentially to the sight of faces. *Science*, **236**: 448–450.

Kendrick, K. M., Atkins, K., Hinton, M. R. et al. (1995). Facial and vocal discrimination in sheep. *Animal Behaviour*, **49**: 1665–1676.

Kendrick, K. M., Atkins, K., Hinton, M. R., Heavens, P. & Keverne, B. (1996). Are faces special for sheep? Evidence from facial and object discrimination learning tests showing effects of inversion and social familiarity. *Behavioural Processes*, **38**: 19–35.

Kendrick, K. M., da Costa, A. P., Leigh, A. E., Hinton, M. R. & Peirce, J. W. (2001). Sheep don't forget a face. *Nature*, **414**: 165–166.

Kight, S. L., Steelman, L., Coffey, G., Lucente, J. & Castillo, M. (2008). Evidence of population level in giant water bugs, *Belostoma flumineum Say* (Heteroptera: Belostomatidae): T-maze turning is left biased. *Behavioural Processes*, **79**: 66–69.

Kilpatrick, L. A., Zald, D. H., Pardo, J. V. et al. (2006). Sex-related differences in amygdala functional connectivity during resting conditions. *NeuroImage*, **30**: 452–461.

Kim, D., Carlson, J. N., Seegal, R. F. & Lawrence, D. A. (1999). Differential immune responses in mice with left and right-turning preference. *Journal of Neuroimmunology*, **93**: 164–171.

Kim, H. (2011). Neural activity that predicts subsequent memory and forgetting: A meta-analysis of 74 fMRI studies. *NeuroImage*, **54**: 2446-2461.

Kimura, D. (1999). *Sex and Cognition*. Cambridge, MA: MIT Press.

King, J. E. & Landau, V. I. (1993). Manual preferences in varieties of reaching in squirrel monkeys. In: J. P. Ward & W. D. Hopkins (eds.), *Primate Laterality: Current Evidence of Primate Asymmetries*, New York: Springer-Verlag, pp. 107-124.

Kipper, S. & Todt, D. (2002). The use of vocal signals in the social play of Barbary Macaques. *Primates*, **43**: 3-17.

Kiuchi, M., Nagata, N., Ikeno, S. & Terakawa, N. (2000). The relationship between the response to external light stimulation and behavioral state in the human fetus: How it differs from vibroacoustic stimulation. *Early Human Development*, **58**: 153-165.

Klein, R. M. & Andrew, R. J. (1986). Distractions, decisions and persistence in runway tests using the domestic chick. *Behaviour*, **99**: 139-156.

Knecht, S., Drager, B., Deppe, M. *et al.* (2000). Handedness and hemispheric language dominance in healthy humans. *Brain*, **123**: 2512-2518.

Knickmeyer, R. C. & Baron-Cohen, S. (2006). Foetal testosterone and sex difference in typical social development and in autism. *Journal of Child Neurology*, **21**: 825-845.

Koboroff, A., Kaplan, G. & Rogers, L. J. (2008). Hemispheric specialization in Australian magpies (*Gymnorhina tibicen*) shown as eye preferences during response to a predator. *Brain Research Bulletin*, **76**: 304-306.

Kocot, K. M., Cannon, J. T., Todt, C. *et al.* (2011). Phylogenomics reveals deep molluscan relationships. *Nature*, **477**: 452-456.

Koechlin, E. & Hyafil, A. (2007). Anterior prefrontal function and the limits of human decision-making. *Science*, **318**: 594-598.

Koechlin, E. & Summerfield, C. (2007). An information theoretical approach to prefrontal executive function. *Trends in Cognitive Science*, **11**: 229-235.

Kon, T., Masahiro, N., Yusuke, Y. *et al.* (2007). Phylogenetic position of a whale-fall lancelet (Cephalochordata) inferred from mitochondrial genome sequences. *BMC Evolutionary Biology*, **7**: 127.

Konkel, A., Warren, D. E., Duff, M. C. *et al.* (2008). Hippocampal amnesia impairs all manner of relational memory. *Frontiers in Human Neuroscience*, **2**. doi:10.3389/neuro.09.015.2008.

Kosslyn, S. M., Chabris, C. F., Marsolek, C. J. & Koenig, O. (1992). Categorical versus coordinate spatial relations: Computational analyses and computer simulations. *Journal of Experimental Psychology: Human Perception and Performance*, **18**: 562-577.

Kovach, J. K. (1968). Spatial orientation of the chick embryo during the last five days of incubation. *Journal of Comparative and Physiological Psychology*, **66**: 283-288.

Krebs, J. R. & Dawkins, R. (1984). Animal signals: Mind reading and manipulation. In: J. R. Krebs & N. B. Davies (eds.), *Behavioural Ecology: An Evolutionary Approach*, Sunderland, MA: Sinauer Associates, pp. 25-50.

Kuan, Y.-S., Gamse, J. T., Schreiber, A. M. & Halpern, M. E. (2007a). Selective asymmetry in a conserved forebrain to midbrain projection. *Journal of Experimental Zoology B: Molecular and Developmental Evolution*, **308**: 669-678.

Kuan, Y.-S., Yu, H.-H., Moens, C. B. & Halpern, M. E. (2007b). Neuropilin asymmetry mediates a left-right difference in habenular connectivity. *Development*, **134**: 857-865.

Kwok, V., Niu, Z., Kay, P. *et al.* (2011). Learning new color names produces rapid increase in gray matter in the intact human cortex. *Proceedings of the National Academy of Sciences USA*, **108**: 6686–6688.

Lacalli, T. C. (1994). Apical organs, epithelial domains and the origin of the chordate central nervous system. *American Zoologist*, **34**: 533–541.

Lacalli, T. C. (1996). Frontal eye circuitry, rostral sensory pathways and brain organisation in amphioxus larvae: Evidence from 3D reconstruction. *Philosophical Transactions of the Royal Society of London B*, **351**: 243–263.

Lacalli, T. C. (2002). The dorsal compartment locomotory control system in amphioxus larvae. *Journal of Morphology*, **252**: 227–237.

Lacalli, T. C. (2008). Basic features of the ancestral chordate brain: A protochordate perspective. *Brain Research Bulletin*, **75**: 319–323.

Lacalli, T. C., Holland, N. D. & West, J. E. (1994). Landmarks in the anterior nervous system of amphioxus larvae. *Philosophical Transactions of the Royal Society of London B*, **344**: 165–185.

LaMendola, N. P. & Bever, T. G. (1997). Peripheral and cerebral asymmetries in the rat. *Science*, **278**: 483–486.

Land, M. F. & Fernald, R. D. (1992). The evolution of eyes. *Annual Review of Neuroscience*, **15**: 1–29.

Lane, R. D. & Jennings, J. R. (1995). Hemispheric asymmetry, autonomic asymmetry and the problem of sudden cardiac death. In: R. J. Davidson & K. Hugdahl (eds.), *Brain Asymmetry*, Cambridge, MA: MIT Press, pp. 271–304.

Langford, D. J., Crager, S. E., Shehzad, Z. *et al.* (2006). Social modulation of pain as evidence for empathy in mice. *Science*, **312**: 1967–1970.

Larose, C., Rogers, L. J., Ricard-Yris, M.-A. & Hausberger, M. (2006). Laterality of horses associated with emotionality in novel situations. *Laterality*, **11**: 355–367.

Lartillot, N. & Philippe, H. (2008). Improvement of molecular phylogenetic inference and the phylogeny of Bilateria. *Philosophical Transactions of the Royal Society of London B*, **363**: 1463–1472.

Laviola, G., Maerí, S., Morley-Fletcher, S. *et al.* (2003). Risk-taking in adolescent mice: Psychobiological determinants and early epigenetic influences. *Neuroscience and Biobehavioral Reviews*, **27**: 19–31.

Lavrysen, A., Heremans, E., Peeters, R. *et al.* (2008). Hemispheric asymmetries in eye–hand coordination. *NeuroImage*, **39**: 1938–1949.

Leavens, D. A., Hopkins, W. D. & Thomas, R. K. (2004). Referential communication by chimpanzees (*Pan troglodytes*). *Journal of Comparative Psychology*, **118**: 48–57.

Leith, J. L., Koutsikou, S., Lumb, B. M. *et al.* (2010). Spinal processing of noxious and innocuous cold information: Differential modulation by the periaqueductal grey. *Journal of Neuroscience*, **30**: 4933–4942.

Letzkus, P., Boeddeker, N., Wood, J. T., Zhang, S. W. & Srinivasan, M. V. (2007). Lateralization of visual learning in the honeybee. *Biology Letters*, **4**: 16–18.

Letzkus, P., Ribi, W. A., Wood, J. T. *et al.* (2006). Lateralisation of olfaction in the honeybee *Apis mellifera*. *Current Biology*, **16**: 1471–1476.

Levin, M., Johnson, R. L., Sten, C. D., Kuehn, M. & Tabin, C. (1995). A molecular pathway determining left–right asymmetry in chick embryogenesis. *Cell*, **82**: 803–814.

Levy, J. (1969). Possible basis for the evolution of lateral specialization of the human brain. *Nature*, **224**: 614–615.

Levy, J., Trevarthen, C. & Sperry, R. W. (1972). Perception of bilateral chimeric figures following hemispheric disconnection. *Brain*, **95**: 61–78.

Li, C., Tierney, C., Wen, L. *et al.* (1997). A single morphogenetic field gives rise to two retina primordial under the influence of the prechordal plate. *Development*, **124**: 603–615.

Lippolis, G., Bisazza, A., Rogers, J. & Vallortigara, G. (2002). Lateralization of predator avoidance responses in three species of toads. *Laterality*, **7**: 163–183.

Lippolis, G., Joss, J. & Rogers, L. J. (2009). Australian lungfish (*Neoceratodus forsteri*): A missing link in the evolution of complementary side biases for predator avoidance and prey capture. *Brain Behavior and Evolution*, **73**: 295–303.

Lippolis, G., Westman, W., McAllan, B. M. & Rogers, L. J. (2005). Lateralization of escape responses in the striped-faced dunnart, *Sminthopsis macroura* (Dasyuridae: Marsupalia). *Laterality*, **10**: 457–470.

Llorente, M., Mosquera, M. & Fabre, M. (2009). Manual laterality for simple reaching and bimanual coordinated task in naturalistic housed chimpanzees (*Pan troglodytes*). *International Journal of Primatology*, **30**: 183–197.

Llorente, M., Riba, D., Palou, L. *et al.* (2011). Population-level right-handedness for a coordinated bimanual task in naturalistic housed chimpanzees: Replication and extension in 114 animals from Zambia and Spain. *American Journal of Primatology*, **73**: 281–290.

Lonsdorf, E. V. & Hopkins, W. D. (2005). Wild chimpanzees show population-level handedness for tool use. *Proceeding of the National Academy of Sciences USA*, **102**: 12634–12638.

Lössner, B. & Rose, S. P. (1983). Passive avoidance training increases fucokinase activity in right forebrain base of day-old chicks. *Journal of Neurochemistry*, **41**: 1357–1363.

Louis, M., Huber, T., Benton, R. *et al.* (2008). Bilateral olfactory sensory input enhances chemotaxis behaviour. *Nature Neuroscience*, **11**: 187–199.

Love, O. P., Wynne-Edwards, K. E., Bond, L. & William, T. D. (2008). Determinants of within- and among-clutch variation in yolk corticosterone in the European starling. *Hormones and Behavior*, **53**: 104–111.

Lowe, C. J., Wu, M., Salic, A. *et al.* (2003). Anteroposterior patterning in hemichordates and the origin of the chordate nervous system. *Cell*, **113**: 853–865.

Lüders, E., Narr, K. L., Thompson, P. M. *et al.* (2006). Hemispheric asymmetries in cortical thickness. *Cerebral Cortex*, **16**: 1232–1238.

Lurito, J. T. & Dzemidzic, M. (2001). Determination of cerebral hemisphere language dominance with functional magnetic resonance imaging. *Neuroimaging Clinics of North America*, **11**: 355–363.

MacNeilage, P. F. (1998). Towards a unified view of cerebral hemispheric specialisations in vertebrates. In: A. Milner (ed.), *Comparative Neuropsychology*, Oxford: Oxford University Press, pp. 167–183.

MacNeilage, P. F. (2007). Present status of the postural origins theory. In: W. D. Hopkins (ed.), *The Evolution of Hemispheric Specialization in Primates*. Special Topics in Primatology, Vol. 5, London: Academic Press, pp. 59–91.

MacNeilage, P. F. (2008). *The Origin of Speech*. Oxford: Oxford University Press.

MacNeilage, P. F., Rogers, L. J. & Vallortigara, G. (2009). Origins of the left and right brain. *Scientific American*, **301**: 60–67.

MacNeilage, P. F., Studdert-Kennedy, M. J. & Lindblom, B. (1987). Primate handedness reconsidered. *Behavioral and Brain Sciences*, **10**: 247–303.

MacPherson, S. E., Turner, M. S., Bozzali, M. *et al.* (2010). Frontal subregions mediating elevator counting task performance. *Neuropsychologia*, **48**: 3679–3682.

Magat, M. & Brown, C. (2009). Laterality enhances cognition in Australian parrots. *Proceedings of the Royal Society of London B*, **276**: 4155–4162.

Maguire, E. A. & Mummery, C. J. (1999). Differential modulation of a common memory retrieval network revealed by positron emission tomography. *Hippocampus*, **9**: 54–61.

Maguire, E. A., Gadian, D. G., Johnsrude, I. S. *et al.* (2000). Navigation-related structural change in the hippocampi of taxi drivers. *Proceedings of the National Academy of Sciences USA*, **97**: 4398–4403.

Maguire, E. A., Woollett, K. & Spiers, H. J. (2006). London taxi drivers and bus drivers: A structural MRI and neuropsychological analysis. *Hippocampus*, **16**: 1091–1101.

Maillard, L., Barbeau, E. J., Baumann, C. *et al.* (2011). From perception to recognition memory: Time course and lateralisation of neural substrates of word and abstract picture processing. *Journal of Cognitive Neuroscience*, **23**: 782–800.

Malaschichev, Y. B. & Wassersug, R. J. (2004). Left and right in the amphibian world: Which way to develop and where to turn? *BioEssays*, **26**: 1–11.

Mallatt, J. (1985). Reconstructing the life cycle and the feeding of ancestral vertebrates. In: R. E. Foreman, A. Gorbman, J. M. Dodd & R. Olsson (eds.), *The Evolutionary Biology of Primitive Fishes*, New York: Plenum Press, pp. 59–68.

Mallatt, J. & Chen, J.-Y. (2003). Fossil sister group of craniates: Predicted and found. *Journal of Morphology*, **258**: 1–31.

Manns, M. & Güntürkün, O. (2009). Dual coding of visual asymmetries in the pigeon brain: The interaction of bottom-up and top-down systems. *Experimental Brain Research*, **199**: 323–332.

Marchant, L. F. & Steklis, H. D. (1986). Hand preference in a captive island group of chimpanzees (*Pan troglodytes*). *American Journal of Primatology*, **10**: 301–313.

Marchant, L. F., McGrew, W. C. & Eibl-Eibesfeldt, I. (1995). Is human handedness universal? Ethological analyses from three traditional cultures. *Ethology*, **101**: 239–258.

Mari, M., Castiello, U., Marks, D. *et al.* (2003). The reach-to-grasp movement in children with autism spectrum disorder. *Philosophical Transactions of the Royal Society of London B*, **358**: 393–403.

Marshall, A. J., Wrangham, R. W. & Arcadi, A. C. (1999). Does learning affect the structure of vocalisations in chimpanzees? *Animal Behaviour*, **58**: 825–830.

Martin, B., Andreas, S., Ramona, K. *et al.* (2010b). Von Economo neuron density in the anterior cingulate cortex is reduced in early onset schizophrenia. *Acta Neuropathologica*, **119**: 771–778.

Martin, F. & Niemitz, C. (2003). 'Right-trunkers' and 'left-trunkers': Side preferences of trunk movements in wild Asian elephants (*Elephas maximus*). *Journal of Comparative Psychology*, **117**: 371–379.

Martin, G. N. & Gray, C. D. (1996). The effect of audience laughter on men's and women's response to humour. *Journal of Social Psychology*, **136**: 221–231.

Martin, G. R. (2009). What is binocular vision for? A birds' eye view. *Journal of Vision*, 9: 1–19.

Martin, J., López, P., Bonati, B. & Csermely, D. (2010a). Lateralization when monitoring predators in the wild: A left eye control of the common wall lizard (*Podarcis muralis*). *Ethology*, 116: 1226–1233.

Marzoli, D. & Tommasi, L. (2009). Side biases in humans (*Homo sapiens*): Three ecological studies on hemispheric asymmetries. *Naturwissenschaften*, 96: 1099–1106.

Matsuo, R., Kawaguchi, E., Yamagishi, M. *et al.* (2010). Unilateral storage in the procerebrum of the terrestrial slug *Limax*. *Neurobiology of Learning and Memory*, 93: 337–342.

Matsusaka, T. (2004). When does play panting occur during social play in wild chimpanzees? *Primates*, 45: 221–228.

Maynard-Smith, J. (1982). *Evolution and the Theory of Games*. Cambridge: Cambridge University Press.

Mazzotti, G. A. & Boere, V. (2009). The right ear but not the left ear temperature is related to stress-induced cortisolaemia in the domestic cat (*Felis catus*). *Laterality*, 14: 196–204.

McCourt, M. E., Garlinglhouse, M. & Butler, J. (2001). The influence of viewing eye on pseudoneglect magnitude. *Journal of the International Neuropsychological Society*, 7: 391–395.

McGilchrist, I. (2009). *The Master and his Emissary*. New Haven, CT: Yale University Press.

McGinnis, M. Y. & Vakulenko, M. (2003). Characterisation of 50-kHz ultrasonic vocalisations in male and female rats. *Physiology and Behaviour*, 80: 81–88.

McGlone, J. (1980). Sex differences in human brain asymmetry. *Behavioral and Brain Sciences*, 3: 215–263.

McGreevy, P. D. & Rogers, L. J. (2005). Motor and sensory laterality in thoroughbred horses. *Applied Animal Behaviour Science*, 92: 337–352.

McGrew, W. C. & Marchant, L. F. (1997). On the other hand: Current issues in and meta-analysis of the behavioral laterality of hand function in nonhuman primates. *American Journal of Physical Anthropology*, 40: 201–232.

McGrew, W. C. & Marchant, L. F. (1999). Laterality of hand use pays off in foraging success for wild chimpanzees. *Primates*, 40: 509–513.

McKenzie, R., Andrew, R. J. & Jones, R. B. (1998). Lateralisation in chicks and hens: New evidence for the control of response by the right eye system. *Neuropsychologia*, 36: 51–58.

McManus, I. C. (1981). Handedness and birth stress. *Psychological Medicine*, 11: 485–496.

McManus, I. C. (1999). Handedness, cerebral lateralization and the evolution of language. In: M. C. Corballis & S. E. G. Lea (eds.), *The Descent of Mind: Psychological Perspectives on Hominid Evolution*, Oxford: Oxford University Press, pp. 194–217.

McManus, I. C. (2002). *Right Hand, Left Hand: The Origins of Asymmetry in Brains, Bodies, Atoms, and Cultures*. London: Weidenfeld & Nicolson.

Meguerditchian, A. & Vauclair, J. (2006). Baboons communicate with their right hand. *Behavioural Brain Research*, 171: 170–174.

Meguerditchian, A., Calcutt, S. E., Lonsdorf, E. V., Ross, S. R. & Hopkins, W. D. (2010a). Captive gorillas are right-handed for bimanual feeding. *American Journal of Physical Anthropology*, 141: 638–645.

Meguerditchian, A., Vauclair, J. & Hopkins, W. D. (2010b). Captive chimpanzees use their right hand to communicate with each other: Implications for the origin of the cerebral substrate for language. *Cortex*, **46**: 40–48.

Mehlhorn, J., Haastert, B. & Rehkämper, G. (2010). Asymmetry of different brain structures in homing pigeons with and without navigational experience. *Journal of Experimental Biology*, **213**: 2219–2224.

Mench, J. & Andrew, R. J. (1986). Lateralisation of a food search task in the domestic chick. *Behavioral and Neural Biology*, **46**: 107–114.

Mendl, M., Burman, O. H. P., Parker, R. M. A. & Paul, E. S. (2009). Cognitive bias as an indicator of animal emotion and welfare: Emerging evidence and underlying mechanisms. *Applied Animal Behaviour Science*, **118**: 161–181.

Merckelbach, H., De Ruiter, C. & Olff, M. (1989). Handedness and anxiety in normal and clinical populations. *Cortex*, **25**: 599–606.

Messenger, J. B. (2001). Cephalopod chromatophores: Neurobiology and natural history. *Biological Reviews*, **76**: 473–528.

Michel, G. F. (1981). Right handedness: A consequence of infant supine head orientation preference? *Science*, **212**: 685–687.

Michel, G. F. & Goodwin, R. (1979). Intrauterine birth position predicts newborn supine head position preference. *Infant Behavior and Development*, **2**: 29–38.

Miklósi, A. & Andrew, R. J. (1999). Right eye use associated with decision to bite in zebrafish. *Behavioural Brain Research*, **105**: 199–205.

Miklósi, A., Andrew, R. J. & Dharmaretnam, M. (1996). Auditory lateralisation: shifts in ear use during attachment in the domestic chick. *Laterality*, **1**: 215–224.

Miklósi, A., Andrew, R. J. & Gasparini, S. (2001). Role of right hemifield in visual control of approach to target in zebrafish. *Behavioural Brain Research*, **106**: 175–180.

Miklósi, A., Andrew, R. J. & Savage, H. (1998). Behavioural lateralisation of the tetrapod type in the zebrafish (*Brachydanio rerio*). *Physiology and Behavior*, **63**: 127–135.

Miller, B. L., Cummings, F., Mishkin, K. *et al.* (1998). Emergence of artistic talent in frontotemporal dementia. *Neurology*, **51**: 978–982.

Miller, M. B., Kingstone, A. & Gazzaniga, M. S. (2002). Hemispheric encoding asymmetry is more apparent than real. *Journal of Cognitive Neuroscience*, **14**: 702–708.

Minagawa-Kawai, Y., Cristia, A. & Dupoux, E. (2011). Cerebral lateralization and early speech acquisition: A developmental scenario. *Developmental Cognitive Neuroscience*, **1**: 217–232.

Miyasaki, N., Morimoto, K., Tsubokawa, T. *et al.* (2009). From the olfactory bulb to higher brain centres: Genetic visualisation of secondary olfactory pathways in zebrafish. *Journal of Neuroscience*, **29**: 4756–4767.

Mobbs, D., Greicius, M. D., Abdel-Azim, E. *et al.* (2003). Humour modulates the mesolimbic reward centres. *Neuron*, **40**: 1041–1048.

Mormann, F., Dubois, J., Kornblith, S. *et al.* (2011). A category-specific response to animals in the right human amygdala. *Nature Neuroscience*, **14**: 1247–1249.

Morris, J. S., Öhman, A. & Dolan, R. J. (1998). Conscious and unconscious emotional learning in the human amygdala. *Nature*, **393**: 467–470.

Morris, J. S., Öhman, A. & Dolan, R. J. (1999). A subcortical pathway to the right amygdala mediating 'unseen' fear. *Proceedings of the National Academy of Sciences USA*, **96**: 1680–1685.

Morris, R. D. & Hopkins, W. D. (1993). Perception of human chimeric faces by chimpanzees: Evidence for a right hemisphere advantage. *Brain and Cognition*, **21**: 111–122.

Mulckhuyse, M. & Theeuwes, J. (2010). Unconscious attentional orienting to exogenous cues: A review of the literature. *Acta Psychologica*, **134**: 299–309.

Mundinger, P. C. (1970). Vocal imitation and individual recognition of finch calls. *Science*, **168**: 480–482.

Myowa-Yamakoshi, M., Tomonaga, M., Tanaka, M. *et al.* (2004). Imitation in neonatal chimpanzees (*Pan troglodytes*). *Developmental Science*, **7**: 437–442.

Nagy, M., Àkos, Z., Biro, D. & Vicsek, T. (2010). Hierarchical group dynamics in pigeon flocks. *Nature*, **464**: 890–894.

Narang, H. K. (1977). Right–left asymmetry of myelin development in epiretinal potion of rabbit optic nerve. *Nature*, **266**: 855–856.

Narang, H. K. & Wisniewski, H. M. (1977). The sequence of myelination in the epiretinal portion of the optic nerve in the rabbit. *Neuropathology and Applied Neurobiology*, **3**: 15–27.

Nepi, M., Cresti, L., Maccagnani, B., Ladurner, E. & Pacini, E. (2005). From the anther to the proctodeum: Pear (*Pyrus communis*) pollen digestion in *Osmia cornuta* larvae. *Journal of Insect Physiology*, **51**: 749–757.

Nestor, P. G. & Safer, M. A. (1990). A multi-method investigation of individual differences in hemisphericity. *Cortex*, **26**: 409–421.

Nicholls, M. E. R., Clode, D., Wood, S. J. & Wood, A. G. (1999). Laterality of expression in portraiture: Putting your best cheek forward. *Proceedings of the Royal Society of London B*, **266**: 1517–1522.

Nixon, M. & Young, J. Z. (2003). *The Brains and Lives of Cephalopods*. Oxford: Oxford University Press.

Nottebohm, F. (1970). Ontogeny of bird song. *Science*, **167**: 950–956.

Nottebohm, F. (1971). Neural lateralization of vocal control in a Passerine bird. I. Song. *Journal of Experimental Zoology*, **177**: 229–261.

Nottebohm, F. (1977). Asymmetries in neural control of vocalization in the canary. In: S. Harnard, R. W. Doty, L. Goldstein, J. Jaynes & G. Krauthamer (eds.), *Lateralization of the Nervous System*, New York: Academic Press, pp. 23–44.

Nottebohm, F. (1980). Brain pathways for vocal learning in birds: A review of the first 10 years. In: J. M. Sprague & A. N. Epstein (eds.), *Progress in Psychobiology and Physiological Psychology*, New York: Academic Press, pp. 85–124.

Nottebohm, F. (1981). Laterality, season and space govern the learning of a motor skill. *Trends in Neuroscience*, **4**: 104–106.

Nottebohm, F., Kasparian, S. & Pandazis, C. (1981). Brain space for a learned task. *Brain Research*, **213**: 99–109.

Nottebohm, F., Stokes, T. M. & Leonard, C. M. (1976). Central control of song in the canary, *Serinus canarius*. *Journal of Comparative Neurology*, **165**: 457–486.

Nowicka, A. & Tacikowski, P. (2011). Transcallosal transfer of information and functional asymmetry of human brain. *Laterality*, **16**: 35–74.

Núñez, J. L. & Juraska, J. M. (1998). The size of the splenium of the rat corpus callosum: Influence of hormones, sex ratio, and neonatal cryoanesthesia. *Developmental Psychobiology*, **33**: 295–303.

Nunn, C. L. (1999). The evolution of exaggerated sexual swellings in primates and the graded signal hypothesis. *Animal Behaviour*, **58**: 229–246.

Ocklenburg, S. & Güntürkün, O. (2009). Head-turning asymmetries during kissing and their association with lateral preference. *Laterality*, **14**: 79–85.

Okubo, M. (2010). Right movies on the right seat: Laterality and seat choice. *Applied Cognitive Psychology*, **42**: 90–99.

Oldfield, R. C. (1971). The assessment and analysis of handedness: The Edinburgh inventory. *Neuropsychologia*, **9**: 97–113.

Olko, C. & Turkewitz, G. (2001). Cerebral asymmetry of emotion and its relationship to olfaction in infancy. *Laterality*, **6**: 29–37.

Ortiz, C. O., Faumont, S., Takayama, J. et al. (2009). Lateralised gustatory behaviour of *C. elegans* is controlled by specific receptor type guanylyl cyclases. *Current Biology*, **19**: 996–1004.

Ott, L., Schleidt, M. & Kien, J. (1994). Temporal organisation of action in baboons: Comparisons with the temporal segmentation in chimpanzee and human behaviour. *Brain, Behaviour and Evolution*, **44**: 101–107.

Palmer, A. R. (1996). Waltzing with asymmetry. *Bioscience*, **46**: 518–532.

Palmer, A. R. (2002). Chimpanzee right-handedness reconsidered: Evaluating the evidence with funnel plots. *American Journal of Physical Anthropology*, **118**: 191–199.

Palmer, A. R. (2003). Reply to Hopkins and Cantalupo: Chimpanzee right-handedness reconsidered. Sampling issues and data presentation. *American Journal of Physical Anthropology*, **121**: 382–384.

Palmer, A. R. (2010). Scale-eating cichlids: From hand(ed) to mouth. *Journal of Biology*, **9**: 11.

Panganiban, G., Irvine, S. M., Lowe, C. et al. (1997). The origin and evolution of animal appendages. *Proceedings of the National Academy of Sciences USA*, **94**: 5162–5166.

Panksepp, J. (2007). Neuroevolutionary sources of laughter and social joy: Modelling primal human laughter in laboratory rats. *Behavioural Brain Research*, **182**: 231–244.

Papadatou, M., Martin, M., Munafo, M. R. et al. (2008). Sex differences in left handedness: A meta-analysis of 144 studies. *Psychological Bulletin*, **134**: 677–699.

Papademetriou, E., Sheu, C. F. & Michel, G. F. (2005). A meta-analysis of primate hand preferences, particularly for reaching. *Journal of Comparative Psychology*, **119**: 33–48.

Park, H. & Rugg, M. D. (2011). Neural correlates of encoding within- and across inter-item associations. *Journal of Cognitive Neuroscience*, **23**: 2533–2543.

Parker, J. D., Keightley, M. L., Smith, C. T. & Taylor, C. J. (1999). Interhemispheric transfer deficit in alexithymia: An experimental study. *Psychosomatic Medicine*, **61**: 464–468.

Parr, L. A. & Hopkins, W. D. (2000). Brain temperature asymmetries and emotional perception in chimpanzees, *Pan troglodytes*. *Physiology and Behavior*, **71**: 363–371.

Pascual, A., Huang, K.-L., Nevue. J. & Préat, T. (2004). Brain asymmetry and long-term memory. *Nature*, **427**: 605–606.

Pasteels, J. J. (1970). Développement embryonnaire. In: P. P. Grassé (ed.), *Traité de Zoologie*, Vol. XIV, Paris: Masson et Cie, pp. 893–971.

Paterson, J. R., Garcia-Bellido, D. C., Lee, M. S. Y. et al. (2011). Acute vision in the giant Cambrian predator *Anomalocaris* and the origin of compount eyes. *Nature*, **480**: 237–240.

Patterson, T. A., Gilbert, D. B. & Rose, S. P. (1990). Pre- and post-training lesions of the intermediate medial hyperstriatum ventrale and passive avoidance learning in the chick. *Experimental Brain Research*, **80**: 189–195.

Pecchia, T., Gagliardo, A., Filannino, C., Ioalè, P. & Vallortigara, G. (2012). Navigating through an asymmetrical brain: Lateralisation and homing in pigeon. In: D. Csermely & L. Regolin (eds.), *Behavioural Lateralization in Vertebrates: Two Sides of the Same Coin*, Berlin, Heidelberg: Springer, in press.

Peirce, J. W. & Kendrick, K. M. (2002). Functional asymmetry in sheep temporal cortex. *NeuroReport*, 13: 2395–2399.

Peirce, J. W., Leigh, A. E., da Costa, A. P. C. & Kendrick, K. M. (2001). Human face recognition in sheep: Lack of configurational coding and right hemisphere advantage. *Behavioural Processes*, 55: 13–26.

Peirce, J. W., Leigh, A. E. & Kendrick, K. M. (2000). Configurational coding, familiarity and the right hemisphere advantage for face recognition in sheep. *Neuropsychologia*, 38: 475–483.

Pellis, S. M. (1981). A description of social play by the Australian magpie *Gymnorhina tibicen* based on Eshkol–Wachman notation. *Bird Behaviour*, 3: 61–79.

Pepperberg, I. M. (1994). Vocal learning in gray parrots (*Psittacus erithacus*) – effects of social interaction, reference and context. *The Auk*, 111: 300–313.

Perelle, I. B. & Ehrman, L. (1994). An international study of human handedness: The data. *Behavioral Genetics*, 24: 217–227.

Peters, H. & Rogers, L. J. (2007). Limb use and preferences in wild orang-utans during feeding and locomotor behavior. *American Journal of Primatology*, 69: 1–15.

Petkov, C. L., Kayser, C., Steudel, T. *et al.* (2008). A voice region in the monkey brain. *Nature Neuroscience*, 11: 367–374.

Pfannkuche, K. A., Bouma, A. & Groothuis, T. G. G. (2009). Does testosterone affect lateralization of brain and behaviour? A meta-analysis in humans and other animal species. *Philosophical Transactions of the Royal Society of London B*, 364: 929–942.

Phelps, E. A., O'Connor, K. J., Gatenby, G. *et al.* (2001). Activation of the left amygdala to a cognitive representation of fear. *Nature Neuroscience*, 4, 437–441.

Phillips, R. E. & Youngren, O. M. (1986). Unilateral kainic acid lesions reveal dominance of right archistriatum in avian fear behaviour. *Brain Research*, 377: 216–220.

Piekema, C., Kessels, R. P. C., Mars, K. B. *et al.* (2006). The right hippocampus participates in short-term memory maintenance of object-location associations. *NeuroImage*, 33: 374–382.

Pierson, J. M., Bradshaw, J. L. & Nettleton, N. C. (1983). Head and body space to left and right, front and rear. 1. Unidirectional competitive auditory stimulation. *Neuropsychologia*, 21: 463–473.

Pilcher, D. L., Hammock, E. A. D. & Hopkins, W. D. (2001). Cerebral volumetric asymmetries in non-human primates: A magnetic resonance imaging study. *Laterality*, 6: 165–179.

Pinsk, M. A., DeSimone, K., Moore, T., Gross, C. G. & Kastner, S. (2005). Representations of faces and body parts in macaque temporal cortex: A functional MRI study. *Proceedings of the National Academy of Sciences USA*, 102: 6996–7001.

Pisella, L., Alahyane, N., Blangero, A. *et al.* (2011). Right hemispheric dominance for visual remapping in humans. *Philosophical Transactions of the Royal Society of London B*, 365: 572–585.

Pizzamiglio, L. & Mammucari, A. (1985). Evidence for sex differences in brain organisation in patients with recovery in aphasia. *Brain and Language*, 25: 213–223.

Pizzamiglio, L., Guariglia, C. & Cosentino, T. (1998). Evidence for separate allocentric and egocentric space processing in neglect patients. *Cortex*, **34**: 719–730.

Poirier, C., Boumans, T., Verhoye, M., Balthazart, J. & Van Der Linden, A. (2009). Own song recognition in the songbird auditory pathway: Selectivity and lateralization. *Journal of Neuroscience*, **29**: 2252–2258.

Poremba, A. & Mishkin, M. (2007). Exploring the extent and function of higher-order auditory cortex in rhesus monkeys. *Hearing Research*, **229**: 14–23.

Prather, J. F., Peters, S., Nowicki, S. *et al.* (2008). Precise-auditory mirroring in neurons for learned vocal communication. *Nature*, **451**: 305–310.

Previc, F. H. (1991). A general theory concerning the prenatal origins of cerebral lateralization in humans. *Psychological Review*, **98**: 299–334.

Proverbio, A. M., Riva, F., Martin, E. *et al.* (2010). Face coding is bilateral in the female brain. *PLoS One*, **5**: e11242.

Pu, G. A. & Dowling, J. E. (1981). Anatomical and physiological characteristics of pineal photoreceptor cells in the larval lamprey, *Petromyzon fluviatilis*. *Journal of Neurophysiology*, **46**: 1018–1038.

Putnam, M. C., Steven, M. S., Doron, K. W. *et al.* (2010). Cortical projection topography of the human splenium: hemispheric asymmetry and individual differences. *Journal of Cognitive Neuroscience*, **22**: 1662–1669.

Puzdrowski, R. L. & Gruber, S. (2009). Morphological features of the cerebellum of the Atlantic stingray and their possible evolutionary significance. *Integrative Zoology*, **4**: 110–122.

Quaranta, A., Siniscalchi, M., Albrizio, M. *et al.* (2008). Influence of behavioural lateralization on interleukin-2 and interleukin-6 gene expression in dogs before and after immunization with rabies vaccine. *Behavioural Brain Research*, **186**: 256–260.

Quaranta, A., Siniscalchi, M., Frate, A. & Vallortigara, G. (2004). Paw preference in dogs: Relations between lateralised behaviour and immunity. *Behavioural Brain Research*, **153**: 521–525.

Quaranta, A., Siniscalchi, M. & Vallortigara, G. (2007). Asymmetric tail-wagging responses by dogs to different emotive stimuli. *Current Biology*, **17**: 199–201.

Rahman, Q. & Wilson, G. D. (2003). Large sexual-orientation-related differences in performance of mental rotation and judgement of line orientation tasks. *Neuropsychology*, **17**: 25–31.

Rahman, Q., Abrahams, S. & Wilson, G. D. (2003). Sexual-orientation-related difference in verbal fluency. *Neuropsychology*, **17**: 240–246.

Raichle, M. E., MacLeod, A. M., Snyder, A. Z. *et al.* (2001). A default mode of brain function. *Proceedings of the National Academy of Sciences USA*, **98**: 676–682.

Rajendra, S. & Rogers, L. J. (1993). Asymmetry is present in the thalamofugal projections of female chicks. *Experimental Brain Research*, **92**: 542–544.

Rashid, N. Y. & Andrew, R. J. (1989). Right hemisphere advantage for topographical orientation in the domestic chick. *Neuropsychologia*, **7**: 937–948.

Raymond, M. & Pontier, D. (2004). Is there geographical variation in human handedness? *Laterality*, **9**: 35–51.

Raymond, M., Pontier, D., Dufour, A. & Moller, A. P. (1996). Frequency-dependent maintenance of left handedness in humans. *Proceedings of the Royal Society of London B*, **263**: 1627–1633.

Regan, J. C., Concha, M. L., Roussigne, M., Russell, C. & Wilson, S. W. (2009). An Fgf8-dependent bistable cell migratory event establishes CNS asymmetry. *Neuron*, **61**: 27–34.

Regier, T., Kay, P. & Khetarpal, N. (2007). Colour naming reflects optimal partitions of colour space. *Proceedings of the National Academy of Sciences USA*, **104**: 1436–1441.

Regolin, L., Marconato. F. & Vallortigara, G. (2004). Hemispheric differences in the recognition of partly occluded objects by newly hatched domestic chicks (*Gallus gallus*). *Animal Cognition*, **7**: 162–170.

Ren, P., Nicholls, M. E. R., Ma, Y.-Y. *et al.* (2011). Size matters: Non-numerical magnitude affects the spatial coding of response. *PLoS One*, **6**: e23553.

Reverberi, C., Shallice, T., D'Agostini, S. *et al.* (2009). Cortical bases of elementary deductive reasoning: inference, memory and metadeduction. *Neuropsychologia*, **47**: 1107–1116.

Reynolds Losin, E. A., Russell, J. L., Freeman, H. *et al.* (2008). Left hemisphere specialisation for oro-facial movements of learned vocal signals by captive chimpanzees. *PLoS One*, **3**: e2529.

Rickard, N. S. & Gibbs, M. E. (2003a). Effects of nitric oxide inhibition on avoidance learning in the chick are lateralized and localized. *Neurobiology of Learning and Memory*, **79**: 252–256.

Rickard, N. S. & Gibbs, M. E. (2003b). Hemispheric dissociation of the involvement of NOS isoforms in memory for discriminated avoidance in the chick. *Learning and Memory*, **10**: 314–318.

Rieger, V., Perez, Y., Muellin, C. H. G. *et al.* (2011). Development of the nervous system in hatchlings of *Spadella cephaloptera* (Chaetognatha) and implications for nervous system evolution in Bilateria. *Development Growth and Differentiation*, **53**: 740–759.

Rigosi, E., Frasnelli, E., Vinegoni, C. *et al.* (2011). Searching for anatomical correlates of olfactory lateralization in the honeybee antennal lobes: A morphological and behavioural study. *Behavioural Brain Research*, **221**: 290–294.

Rizzolatti, G. & Arbib, M. A. (1998). Language within our grasp. *Trends in Neurosciences*, **21**: 188–194.

Rizzolatti, G., Fadiga, L., Gallese, V. & Fogassi, L. (1996). Premotor cortex and the recognition of motor actions. *Cognitive Brain Research*, **3**: 131–141.

Roberson, D. & Hanley, J. R. (2009). Only half right: Comment on Regier and Kay. *Trends in Cognitive Sciences*, **13**: 500.

Robert, M. & Ohlman, T. (1994). Water-level representation by men and women as a function of rod-and-frame test proficiency and visual and postural information. *Perception*, **23**: 1321–1333.

Roberts, A. (1978). Pineal eye and behaviour in *Xenopus* tadpoles. *Nature*, **273**: 774–775.

Robins, A. & Phillips, C. (2010). Lateralised visual processing in domestic cattle herds responding to novel and familiar stimuli. *Laterality*, **15**: 514–534.

Robins, A. & Rogers, L. J. (2004). Lateralised prey catching responses in the toad (*Bufo marinus*): Analysis of complex visual stimuli. *Animal Behaviour*, **68**: 567–575.

Robins, R. & Rogers, L. J. (2006). Complementary and lateralized forms of processing in *Bufo marinus* for novel and familiar prey. *Neurobiology of Learning and Memory*, **86**: 214–227.

Robins, A., Lippolis, G., Bisazza, A., Vallortigara, G. & Rogers, L. J. (1998). Lateralized agonistic responses and hindlimb use in toads. *Animal Behaviour*, **56**: 875–881.

Robinson, R. G. (1985). Lateralized behavioural and neurochemicaal consequences of unilateral brain injury in rats. In: S. D. Glick (ed.), *Cerebral Lateralization in Nonhuman Species*, Orlando, FL: Academic Press, pp. 135–156.

Robinson, R. G. & Downhill, P. (1995). Lateralization of psychopathology in response to focal brain injury. In: R. J. Davidson & K. Hugdahl (eds.), *Brain Asymmetry*, London: MIT Press, pp. 693–711.

Robinson, R. G., Kubos, K. L., Starr, L. B. *et al.* (1984). Mood disorders in stroke patients: Importance of location of lesion. *Brain*, **107**: 81–93.

Rodriguez, F., Lopez, J. C., Vargas, J. P. *et al.* (2002). Spatial memory and hippocampal pallium through vertebrate evolution: Insights from reptiles and teleost fish. *Brain Research Bulletin*, **57**: 499–503.

Rogers, L. J. (1974). Persistence and search influenced by natural levels of androgens in young and adult chickens. *Physiology and Behavior*, **12**: 197–204.

Rogers, L. J. (1980). Lateralisation in the avian brain. *Bird Behaviour*, **2**: 1–12.

Rogers, L. J. (1982). Light experience and asymmetry of brain function in chickens. *Nature*, **297**: 223–225.

Rogers, L. J. (1990). Light input and the reversal of functional lateralization in the chicken brain. *Behavioural Brain Research*, **38**: 211–221.

Rogers, L. J. (1991). Development of lateralisation. In: R. J. Andrew (ed.), *Neural and Behavioural Plasticity: The Use of the Domestic Chicken as a Model*, Oxford: Oxford University Press, pp. 507–535.

Rogers, L. J. (1995). *The Development of Brain and Behaviour in the Chicken*. Wallingford: CAB International:.

Rogers, L. J. (1999a). Effect of light exposure of eggs on posthatching behaviour of chickens. In: N. Adams & R. Slotow (eds.), *Making Rain for African Ornithology*. Proceedings of the 22nd International Ornithological Congress 16–22 August 1998, Durban. Johannesburg: Birdlife South Africa, S46.2.

Rogers, L. J. (1999b). *Sexing the Brain*. London: Weidenfeld and Nicolson.

Rogers, L. J. (2000). Evolution of hemispheric specialisation: Advantages and disadvantages. *Brain and Language*, **73**: 236–253.

Rogers, L. J. (2002a). Lateralization in vertebrates: Its early evolution, general pattern and development. In: P. J. B. Slater, J. Rosenblatt, C. Snowdon & T. Roper (eds.), *Advances in the Study of Behavior*, Vol. 31, San Diego, CA: Academic Press, pp. 107–162.

Rogers, L. J. (2002b). Advantages and disadvantages of lateralization. In: L. J. Rogers & R. J. Andrew (eds.), *Comparative Vertebrate Lateralization*, Cambridge: Cambridge University Press, pp. 126–153.

Rogers, L. J. (2002c). Lateralized brain function in anurans: Comparison to lateralization in other vertebrates. *Laterality*, **7**: 219–239.

Rogers, L. J. (2006). Cognitive and social advantages of a lateralized brain. In: Y. B. Malashichev & A. W. Deckel (eds.), *Behavioral and Morphological Asymmetries in Vertebrates*, Texas: Landes Bioscience, pp. 129–139.

Rogers, L. J. (2008). Development and function of lateralization in the avian brain. *Brain Research Bulletin*, **76**: 235–244.

Rogers, L. J. (2009). Hand and paw preferences in relation to the lateralised brain. *Philosophical Transactions of the Royal Society of London B*, **364**: 943–954.

Rogers, L. J. (2010a). Relevance of brain and behavioural lateralization to animal welfare. *Applied Animal Behaviour Science*, **127**: 1–11.

Rogers, L. J. (2010b). Interactive contributions of genes and early experience to behavioural development: Sensitive periods and lateralized brain and behaviour. In: K. E. Hood, C. T. Halpern, G. Greenberg & R. M. Lerner (eds.), *Handbook of Developmental Science, Behavior, and Genetics*, Malden, MA: Wiley-Blackwell Publishing, pp. 400–433.

Rogers, L. J. (2010c). Relevance of brain and behavioural lateralization to animal welfare. *Applied Animal Behaviour Science*, **127**: 1–11.

Rogers, L. J. (2011a). Does brain lateralization have practical implications for improving animal welfare? *CAB Reviews: Perspectives in Agriculture, Veterinary Science, Nutrition and Natural Resources*, **6**(36), 1–10.

Rogers, L. J. (2011b). The two hemispheres of the avian brain: Their differing roles in perceptual processing and the expression of behavior. *Journal of Ornithology*, **153** (Suppl. 1): S61–S74.

Rogers, L. J. & Andrew, R. J. (1989). Frontal and lateral visual field use after treatment with testosterone. *Animal Behaviour*, **38**: 394–405.

Rogers, L. J. & Andrew, R. J. (eds.) (2002). *Comparative Vertebrate Lateralization*. Cambridge: Cambridge University Press.

Rogers, L. J. & Anson, J. M. (1979). Lateralisation of function in the chicken forebrain. *Pharmacology Biochemistry and Behaviour*, **10**: 679–686.

Rogers, L. J. & Bell, G. A. (1989). Different rates of functional development in the two visual systems of the chicken revealed by {$^{14}$C} 2-deoxyglucose. *Developmental Brain Research*, **49**: 161–172.

Rogers, L. J. & Deng, C. (1999). Light experience and lateralization of the two visual pathways in the chick. *Behavioural Brain Research*, **98**: 277–287.

Rogers, L. J. & Deng, C. (2005). Corticosterone treatment of the chick embryo affects light-stimulated development of the thalamofugal visual pathway. *Behavioural Brain Research*, **159**: 63–71.

Rogers, L. J. & Kaplan, G. (1996). Hand preferences and other lateral biases in rehabilitated orang-utans, *Pongo pygmaeus pygmaeus*. *Animal Behaviour*, **51**: 13–25.

Rogers, L. J. & Kaplan, G. (2006). An eye for a predator: Lateralization in birds, with particular reference to the Australian magpie. In: Y. Malashichev & W. Deckel (eds.) *Behavioral and Morphological Asymmetries in Vertebrates*, Texas: Landes Bioscience, pp. 47–57.

Rogers, L. J. & Rajendra, S. (1993). Modulation of the development of light-initiated asymmetry in chick thalamofugal visual projections by oestradiol. *Experimental Brain Research*, **93**: 89–94.

Rogers, L. J. & Sink, H. S. (1988). Transient asymmetry in the projections of the rostral thalamus to the visual hyperstriatum of the chicken, and reversal of its direction by light exposure. *Experimental Brain Research*, **70**: 378–384.

Rogers, L. J. & Vallortigara, G. (2008). From antenna to antenna: Lateral shift of olfactory memory in honeybees. *PLoS One*, **3**: e2340.

Rogers, L. J. & Workman, L. (1989). Light exposure during incubation affects competitive behaviour in domestic chicks. *Applied Animal Behaviour Science*, 23: 187–198.

Rogers, L. J. & Workman, L. (1993). Footedness in birds. *Animal Behaviour*, 45: 409–411.

Rogers, L. J., Andrew, R. J. & Burne, T. H. J. (1998). Light exposure of the embryo and development of behavioural lateralisation in chicks: I. Olfactory responses. *Behavioural Brain Research*, 97: 195–200.

Rogers, L. J., Munro, U., Freire, R., Wiltschko, R. & Wiltschko, W. (2008). Lateralized response of chicks to magnetic cues. *Behavioural Brain Research*, 186: 66–71.

Rogers, L. J., Zappia, J. V. & Bullock, S. P. (1985). Testosterone and eye-brain asymmetry for copulation in chickens. *Experientia*, 41: 1447–1449.

Rogers, L. J., Zucca, P. & Vallortigara, G. (2004). Advantage of having a lateralized brain. *Proceedings of the Royal Society of London B*, 271: S420–S422.

Rosa, C., Lassonde, M., Pinard, C., Keenan, J. P. & Belin, P. (2008). Investigations of hemispheric specialization of self-voice recognition. *Brain Cognition*, 68: 204–214.

Rosa Salva, O., Daisley, J. N., Regolin, L. & Vallortigara, G. (2009). Lateralization of social learning in the domestic chick (*Gallus gallus domesticus*): Learning to avoid. *Animal Behaviour*, 78: 847–856.

Rosa Salva, O., Regolin, L., Mascalzoni, E. & Vallortigara, G. (2012). Cerebral and behavioural asymmetries in animal social recognition. *Comparative Cognition and Behavior Reviews*, in press.

Rose, S. P. (1992). *The Making of Memory*. London: Bantam Press.

Rose, S. P. (2000). God's organism? The chick as a model system for memory studies. *Learning and Memory*, 7: 1–17.

Rota-Stabelli, O., Kayal, E., Gleeson, D. et al. (2010). Ecdysozoan mitogenomics: Evidence for a common origin of the legged invertebrates, the Panarthropoda. *Genome Biology and Evolution*, 2: 425–440.

Roussigné, M., Bianco, I. H, Wilson, S. W. & Blader, P. (2009). Nodal signalling imposes left–right asymmetry upon neurogenesis in the habenular nuclei. *Development*, 136: 1549–1557.

Rowe, T. B., Macrini, T. E. & Luo, Z.-X. (2011). Fossil evidence on origin of mammalian brain. *Science*, 332: 955–957.

Rugani, R., Kelly, D. M., Szelest, I. et al. (2010). Is it only humans that count from left to right? *Biology Letters*, 6: 290–292.

Rugani, R., Vallortigara, G., Vallini, B. & Regolin, L. (2011). Asymmetrical number-space mapping in the avian brain. *Neurobiology of Learning and Memory*, 95: 231–238.

Ryan, B. C. & Vandenbergh, J. G. (2002). Intrauterine position effect. *Neuroscience and Biobehavioral Reviews*, 26: 665–678.

Sackeim, H. A., Weiman, A. L., Gur, R. C. et al. (1982). Pathological laughing and crying: Functional brain asymmetry in the experience of positive and negative emotions. *Archives of Neurology*, 39: 210–218.

Sagasti, A. (2007). Three ways to make two sides: Genetic models of asymmetric nervous system development. *Neuron*, 55: 345–351.

Saint-Galli, A., Marchand, A. R., Decorte, L. et al. (2011). Retrospective evaluation and its neuronal circuit in rats. *Behavioural Brain Research*, 223: 262–270.

Sakai, M., Hishii, T., Takeda, S. & Kohshima, S. (2006). Laterality of flipper rubbing behaviour in wild bottlenose dolphins (*Tursiops aduncus*): Caused by asymmetry of eye use? *Behavioural Brain Research*, 170: 204–210.

Samara, A., Vougas, K., Papadopoulou, A. *et al.* (2011). Proteomics reveal rat hippocampal lateral asymmetry. *Hippocampus*, 21: 108–119.

Sandi, C., Patterson, T. A. & Rose, S. P. (1993). Visual input and lateralization of brain function in learning in the chick. *Neuroscience*, 52: 393–401.

Sandoz, J.-C., Hammer, M. & Menzel, R. (2002). Side-specificity of olfactory learning in the honeybee: US input side. *Learning and Memory*, 9: 337–348.

Santrock, J. W. (2008). Motor, sensory, and perceptual development. In: M. Ryan (ed.), *A Topical Approach to Life-Span Development*, Boston: McGraw-Hill Higher Education, pp. 172–205.

Savic, I. & Lindström, P. (2008). PET and MRI show difference in cerebral asymmetry and functional connectivity between homo- and heterosexual subjects. *Proceedings of the National Academy of Sciences USA*, 105: 9403–9408.

Sayigh, L. S., Esch, H. C., Wells, R. S. *et al.* (2007). Facts about signature whistles of bottlenose dolphins *Tursiops truncatus*. *Animal Behaviour*, 74: 1631–1642.

Saykin, A. J., Johnson, S. C., Flashman, L. A. *et al.* (1999). Functional differentiation of medial temporal and frontal regions involved in processing novel and familiar words: An fMRI study. *Brain*, 122: 1963–1971.

Schaeffel, F., Howland, H. C. & Farkas, L. (1986). Natural accomodation in the growing chicken. *Vision Research*, 26: 1977–1993.

Schenker, N. M., Hopkins, W. D., Spocter, M. A. *et al.* (2010). Broca's area homologue in chimpanzees (*Pan troglodytes*): Probabilistic mapping, asymmetry and comparison to humans. *Cerebral Cortex*, 20: 730–742.

Schiff, B. B. & Lamon, M. (1989). Inducing emotion by unilateral contraction of facial muscles: A new look at hemispheric specialisation and the experience of emotion. *Neuropsychologia*, 27: 923–935.

Schiff, B. B. & Lamon, M. (1994). Inducing emotion by unilateral contraction of hand muscles. *Cortex*, 30: 247–254.

Schmidt, M. F., Ashmore, R. C. & Vu, E. T. (2004). Bilateral control and interhemispheric coordination in the avian song motor system. *Annals of the New York Academy of Sciences*, 1016: 171–186.

Schomerus, C., Korf, H. W., Laedtke, E. *et al.* (2008). Nocturnal behaviour and rhythmic *Period* gene expression in a lancelet, *Branchiostoma lanceolatum*. *Journal of Biological Rhythms*, 23: 170.

Schulte, T. & Müller-Oehring, E. M. (2010). Contribution of callosal connections to the interhemispheric integration of visuomotor and cognitive processes. *Neuropsychological Review*, 20: 174–190.

Schwabl, H. (1999). Developmental changes and among-sibling variation of corticosterone levels in an altricial avian species. *General and Comparative Endocrinology*, 116: 403–408.

Schwarz, I. M. & Rogers, L. J. (1992). Testosterone: A role in the development of brain asymmetry in the chick. *Neuroscience Letters*, 146: 167–170.

Seeck, M., Michel, C. M., Mainwaring, N. *et al.* (1997). Evidence for rapid face recognition from human scalp and intracranial electrodes. *NeuroReport*, 8: 2749–2754.

Seeger, G., Braus, R. F., Kut, M. *et al.* (2002). Body image distortion reveals amygdala activation in patients with anorexia nervosa – a functional magnetic resonance imaging study. *Neuroscience Letters*, **326**: 25–29.

Seeley, W. W., Carlin, D. A. & Allman, J. A. (2006). Early frontotemporal dementia targets neurons unique to apes and humans. *Annals of Neurology*, **60**: 660–667.

Seger, C. A., Poldrack, R. A., Prabhalcaran, V. *et al.* (2000). Hemispheric asymmetries and individual differences in visual concept learning as measured by functional MRI. *Neuropsychologia*, **38**: 1316–1324.

Semendeferi, K., Teffer, K., Buxhoeveden, D. P. *et al.* (2011). Spatial organisation of neurons in the frontal pole sets humans apart from great apes. *Cerebral Cortex*, **21**: 1485–1497.

Shallice, T., Burgess, P. & Robertson, I. (1996). The domain of supervisory processes and temporal organisation of behaviour. *Philosophical Transactions of the Royal Society of London B*, **351**: 1405–1412.

Shamay-Tsoori, S. G., Adler, N., Aharon-Peretz, J. *et al.*(2011). The origins of originality: The neural bases of creative thinking and originality. *Neuropsychologia*, **49**: 178–185.

Shapleski, J., Rossell, S. L., Woodruff, P. W. R. *et al.* (1999). The planum temporale: A systematic, quantitative review of its structural, functional and clinical significance. *Brain Research Review*, **29**: 26–49.

Sharp, P. E., Turner-Williams, S. & Tuttle, S. (2006). Movement-related correlates of single cell activity in the interpeduncular nucleus and habenula of the rat. *Behavioural Brain Research*, **166**: 55–70.

Shaw, J., Claridge, G. & Clark, K. (2001). Schizotypy and the shift from dextrality: A study of handedness in a large non-clinical sample. *Schizophrenia Research*, **50**: 181–189.

Shaywitz, B. A., Shaywitz, S. E., Pugh, K. R. *et al.* (1995). Sex differences in the functional organisation of the brain for language. *Nature*, **373**: 607–609.

Sherry, D. F. & Schachter, D. L. (1987). The evolution of multiple memory systems. *Psychological Review*, **94**: 439–454.

Sherwood, C. C., Duka, T., Simpson, C. D. *et al.* (2010). Neonatal synaptophysin asymmetry and behavioural lateralisation in chimpanzees (*Pan troglodytes*). *European Journal of Neuroscience*, **31**: 1456–1464.

Sherwood, C. C., Wahl, E., Erwin, J. M., Hof, P. R. & Hopkins, W. D. (2007). Histological asymmetries of primary motor cortex predict handedness in chimpanzees (*Pan troglodytes*). *Journal of Comparative Neurology*, **503**: 525–537.

Shettleworth, S. J. (2003). Memory and hippocampal specialization in food-storing birds: Challenges for research on comparative cognition. *Brain Behavior and Evolution*, **62**: 108–116.

Shin, L. M., McNally, R. J., Kosslyn, S. M. *et al.* (1999). Regional cerebral blood flow during script-driven imagery in childhood sexual abuse-related PTSD: A PET investigation. *American Journal of Psychiatry*, **156**: 575–584.

Shinohara, Y., Hosoya, A., Yamasaki, N. *et al.* (2012). Right-hemispheric dominance of spatial memory in split-brain mice. *Hippocampus*, **22**: 117–121.

Sholl, A. A. & Kim, K. L. (1990). Androgen receptors are differentially distributed between right and left cerebral hemispheres of the fetal male rhesus monkey. *Brain Research*, **516**: 122–126.

Shomrat, T., Zarrella, I., Fiorito, G. *et al.* (2008). The octopus vertical lobe modulates short-term learning rate and uses LTP to acquire long-term memory. *Current Biology,* **18**: 337–342.

Shu, D. G., Conway Morris, S., Han, J. *et al.* (2003). Head and backbone of the early Cambrian vertebrate *Haikouichthys. Nature,* **421**: 526–529.

Shulman, G. L., Pope, D. L. W., Astafiev, S. V. *et al.* (2010). Right hemisphere dominance during spatial selective attention and target detection occurs outside the dorsal frontoparietal network. *Journal of Neuroscience,* **30**: 3640–3651.

Siegel, P. B., Isakson, S. T., Coleman, F. N. & Huffman, B. J. (1969). Photoacceleration of development in chick embryos. *Comparative Biochemistry and Physiology,* **28**: 753–758.

Sindhurakar, A. & Bradley, N. S. (2010). Kinematic analysis of overground locomotion in chicks incubated under different light conditions. *Developmental Psychobiology,* **52**: 802–812.

Siniscalchi, M., Dimatteo, S., Pepe, A. M., Sasso, R. & Quaranta, A. (2012). Visual lateralization in wild striped dophins (*Stenella coeruleoalba*) in response to stimuli with different degrees of familiarity. *PloS One,* **7**: e30001.

Siniscalchi, M., Quaranta, A. & Rogers, L. J. (2008). Hemispheric specialization in dogs for processing different acoustic stimuli. *PloS One,* **3**(10): e3349.

Siniscalchi, M., Sasso, R., Pepe, A. M. *et al.* (2010a). Catecholamine plasma levels following immune stimulation with rabies vaccine in dogs selected for their paw preferences. *Neuroscience Letters,* **476**: 142–145.

Siniscalchi, M., Sasso, R., Pepe, A. M. *et al.* (2011). Sniffing with the right nostril: Lateralisation of response to odour stimuli by dogs. *Animal Behaviour,* **82**: 399–404.

Siniscalchi, M., Sasso, R., Pepe, A. M., Vallortigara, G. & Quaranta, A. (2010b). Dogs turn left to emotional stimuli. *Behavioural Brain Research,* **208**: 516–521.

Siok, W. T., Kay, P., Wang, W. S. Y. *et al.* (2009). Language regions of the brain are operative in color perception. *Proceedings of the National Academy of Sciences USA,* **106**: 8140–8145.

Smaers, J. B., Steele, J., Case, C. R. *et al.* (2011). Primate prefrontal cortex evolution: Human brains are at the extreme of a lateralised ape trend. *Brain Behavior and Evolution,* **77**: 67–78.

Smart, J. L., Tonkiss, J. & Massey, R. F. (1986). A phenomenon: Left-biased asymmetrical eye-opening in artificially reared rat pups. *Developmental Brain Research,* **28**: 134–136.

Smith, A. B. (2005). The pre-radial history of echinoderms. *Geological Journal,* **40**: 255–280.

Smotherman, W. P., Brown, C. P. & Levine, S. (1977). Maternal responsiveness following differential pup treatment and mother–pup interactions. *Hormones and Behavior,* **8**: 242–253.

Snyder, A. & Mitchell, D. J. (1999). Is integer arithmetic fundamental to mental processing? The mind's secret arithmetic. *Proceedings of the Royal Society of London B,* **266**: 165–191.

Snyder, A., Brahmanali, H., Hawker, T. & Mitchell, D. J. (2006). Savant-like numerosity skills revealed in normal people by magnetic pulses. *Perception,* **35**: 837–845.

Sovrano, V. A. (2004). Visual lateralization in response to familiar and unfamiliar stimuli in fish. *Behavioural Brain Research,* **152**: 385–391.

Sovrano, V. A. & Andrew, R. J. (2006). Eye use during viewing a reflection: behavioural lateralization in zebrafish larvae. *Behavioural Brain Research*, 167: 226–231.

Sovrano, V. A., Bisazza, A. & Vallortigara, G. (2001). Lateralization of response to social stimuli in fishes: A comparison between different methods and species. *Physiology and Behavior*, 74: 237–244.

Sovrano, V. A., Rainoldi, C., Bisazza, A. & Vallortigara, G. (1999). Roots of brain specializations: Preferential left-eye use during mirror-image inspection in six species of teleost fish. *Behavioural Brain Research*, 106: 175–180.

Spear, L. P. (2000). The adolescent brain and age-related behavioural manifestations. *Neuroscience and Biobehavioral Reviews*, 24: 417–463.

Spivak, B., Segal, M., Mester, R. & Weizman, A. (1998). Lateral preference in post-traumatic stress disorder. *Psychological Medicine*, 28: 229–232.

Spocter, M. A., Hopkins, W. D., Garrison, A. R. *et al.* (2010). Wernicke's area homologue in chimpanzees (*Pan troglodytes*) and its relation to the appearance of modern language. *Proceedings of the Royal Society of London B*, 277: 2165–2174.

Spreng, R. N., Mar, R. A. & Kim, A. S. N. (2008). The common neural basis of autobiographical memory, prospection, navigation, theory of mind, and the default mode: A quantitative meta-analysis. *Journal of Cognitive Neuroscience*, 21: 485–510.

Stewart, T. A. & Albertson, R. C. (2010). Evolution of a unique predatory feeding apparatus: Functional anatomy, development and a genetic locus for jaw laterality in Lake Tanganyika scale-eating cichlids. *BMC Biology*, 8: 8.

Stokes, M. D. (1997). Larval locomotion of the lancelet *Branchiostoma floridae*. *Journal of Experimental Biology*, 200: 1661–1680.

Stokes, M. D. & Holland, D. D. (1995). Ciliary hovering in larval lancelets. *Biological Bulletin*, 188: 231–233.

Stoodley, C. J. & Schmahmann, J. D. (2009). Functional tomography in the human cerebellum: A meta-analysis on neuroimaging studies, *NeuroImage*, 44: 489–501.

Sullivan, R. M. (2004). Hemispheric asymmetry in stress processing in rat prefrontal cortex and the role of mesocortical dopamine. *Stress*, 7: 131–143.

Summers, M. J., Crowe, S. F. & Ng, K. T. (1996). Administration of lanthanum chloride following a reminder induces transient loss of memory in the day-old chick. *Cognitive Brain Research*, 4: 109–119.

Sutherland, R. J. (1982). The dorsal diencephalic conduction system: A review of the anatomy and function of the habenular complex. *Neuroscience Biobehavioral Reviews*, 6: 1–13.

Suthers, R. A. (1990). Contributions to birdsong from the left and right sides of the intact syrinx. *Nature*, 347: 473–477.

Swalla, B. J. & Smith, A. B. (2008). Deciphering deuterostome phylogeny: Molecular, morphological and palaeontological perspectives. *Philosophical Transactions of the Royal Society of London B*, 363: 1557–1568.

Szaniawski, H. (2009). The earliest known venomous animals recognised amongst conodonts. *Acta Palaeontologica Polonica*, 54: 669–676.

Tager-Flusberg, H. & Joseph, R. M. (2003). Identifying neurocognitive phenotypes in autism. *Philosophical Transactions of the Royal Society of London B*, 358: 303–314.

Taglialatela, J. P., Russell, J. L., Schaeffer, J. A. *et al.* (2011). Chimpanzee vocal signalling points to a multimodal origin of human language. *PLoS One*, 6: e18852.

Tan, U. (1987). Paw preferences in dogs. *International Journal of Neuroscience*, 32: 825–829.

Tang, A. C. & Reeb, B. C. (2003). Neonatal novelty exposure, dynamics of brain asymmetry, and social recognition memory. *Developmental Psychobiology*, 44: 84–93.

Tang, A. C. & Verstynen, T. (2002). Early life environment modulates 'handedness' in rats. *Behavioural Brain Research*, 131: 1–7.

Taylor, R. W., Hsieh, Y. W., Gamse, J. T. & Chuang, C. F. (2010). Making a difference together: Reciprocal interactions in *C. elegans* and zebrafish asymmetric neural development. *Development*, 137: 681–691.

Telford, M. J., Bourlat, S. J., Economou, A. *et al.* (2008). The evolution of the Ecdysozoa. *Philosophical Transactions of the Royal Society of London B*, 218: 333–339.

Tennie, C., Hedwig, D., Call, J. & Tomasello, M. (2008). An experimental study of nettle feeding in captive gorillas. *American Journal of Primatology*, 70: 584–593.

Teufel, C., Ghazanfar, A. A. & Fischer, J. (2010). On the relationship between lateralised brain function and orienting asymmetries. *Behavioral Neuroscience*, 124: 437–445.

Thatcher, W. W., Walker, R. A. & Giudice, S. (1987). Human cerebral hemispheres develop at different rates and ages. *Science*, 236: 1110–1113.

Thomas, P. O. R., Croft, D. P., Marshall, L. J. *et al.* (2008). Does defection during predator inspection affect social structure in wild shoals of guppies? *Animal Behaviour*, 75: 43–53.

Tomaz, C., Verburg, M. S., Boere, V., Pianta, T. F. & Belo, M. (2003). Evidence of hemispheric specialization in marmosets (*Callithrix penicillata*) using tympanic membrane thermometry. *Brazilian Journal of Medical and Biological Research*, 36: 913–918.

Tomer, R., Denes, A. S., Tessmar-Raible, K. *et al.* (2010). Profiling by image registration reveals common origin of annelid mushroom bodies and vertebrate pallium. *Cell*, 142: 800–809.

Tommasi, L. & Andrew, R. J. (2002). The use of viewing posture to control visual processing by lateralised mechanisms. *Journal of Experimental Biology*, 205: 1451–1457.

Tommasi, L. & Vallortigara, G. (1999). Footedness in binocular and monocular chicks. *Laterality*. 4: 89–95.

Tommasi, L. & Vallortigara, G. (2001). Encoding of geometric and landmark information in the left and right hemispheres of the avian brain. *Behavioral Neuroscience*, 115: 602–613.

Tommasi, L. & Vallortigara, G. (2004). Hemispheric processing of landmark and geometric information in male and female domestic chicks (*Gallus gallus*). *Behavioural Brain Research*, 155: 85–96.

Tommasi, L., Andrew, R. J. & Vallortigara, G. (2000). Eye use is determined by the nature of task in the domestic chick (*Gallus gallus*). *Behavioral Brain Research*, 112: 119–126.

Tommasi, L., Gagliardo, A., Andrew, R. J. & Vallortigara, G. (2003). Separate processing mechanisms for encoding geometric and landmark information in the avian hippocampus. *European Journal of Neuroscience*, 17: 1695–1702.

Tommasi, L., Vallortigara, G. & Zanforlin, M. (1997). Young chickens learn to localize the centre of a spatial environment. *Journal of Comparative Physiology A: Neuroethology, Sensory, Neural, and Behavioral Physiology*, 180: 567–572.

Tourville, J. A., Reilly, K. J. & Guenther, F. H. (2008). Neural mechanisms underlying auditory feedback control of speech. *NeuroImage*, **39**: 1429–1443.

Town, S. M. (2011). Preliminary evidence of a neurophysiological basis for individual discrimination in filial imprinting. *Behavioural Brain Research*, **225**: 651–654.

Tranel, D., Bechara, A. & Denberg, N. L. (2002). Asymmetric functional roles of right and left ventromedial prefrontal cortices. *Cortex*, **38**: 589–612.

Tsakiris, M. (2010). My body in my brain: A neurocognitive model of body ownership. *Neuropsychologia*, **48**: 703–712.

Tucker, D. M. & Frederick, S. L. (1989). Emotion and brain lateralisation. In: H. Wagner & A. Manstead (eds.), *Handbook of Social Psychophysiology*, Chichester: Wiley, pp. 27–70.

Tully, T. & Quinn, W. G. (1985). Classical conditioning and retention in normal and mutant *Drosophila melanogaster*. *Journal of Comparative Physiology A*, **157**: 263–277.

Tulogdi, A., Toth, M., Halasz, J. *et al.* (2010). Brain mechanisms involved in predatory aggression are activated in a laboratory model of violent intra-specific aggression. *European Journal of Neuroscience*, **32**: 1744–1753.

Tulving, E., Kapur, S., Craik, F. I. *et al.* (1994). Hemispheric encoding/retrieval asymmetry in episodic memory: Positron emission tomography findings. *Proceedings of the National Academy of Sciences USA*, **91**: 2016–2020.

Valencia-Alfonso, C. E., Verhaal, J. & Güntürkün, O. (2009). Ascending and descending mechanisms of visual lateralization in pigeons. *Philosophical Transactions of the Royal Society of London B*, **364**: 955–963.

Valenti, A., Sovrano, V. A., Zucca, P. & Vallortigara, G. (2003). Visual lateralization in quails. *Laterality*, **8**: 67–78.

Vallortigara, G. (1992). Right hemisphere advantage for social recognition in the chick. *Neuropsychologia*, **30**: 761–768.

Vallortigara, G. (2000). Comparative neuropsychology of the dual brain: A stroll through left and right animals' perceptual worlds. *Brain and Language*, **73**: 189–219.

Vallortigara, G. (2004). Visual cognition and representation in birds and primates. In: L. J. Rogers & G. Kaplan (eds.), *Comparative Vertebrate Cognition: Are Primates Superior to Non-Primates?* New York: Kluwer Academic/Plenum Publishers, pp. 57–94.

Vallortigara, G. (2005). Editorial for cortex forum: Cerebral lateralization: A common theme in the organization of the vertebrate brain. *Cortex*, **42**: 5–7.

Vallortigara, G. (2006a). The evolution of behavioural and brain asymmetries: Bridging together neuropsychology and evolutionary biology. In: Y. Malashichev & W. Deckel (eds.), *Behavioral and Morphological Asymmetries in Vertebrates*, Austin, TX: Landes Bioscience, pp. 1–20.

Vallortigara, G. (2006b). The cognitive chicken: Visual and spatial cognition in a non-mammalian brain. In: E. A. Wasserman & T. R. Zentall (eds.), *Comparative Cognition: Experimental Explorations of Animal Intelligence*, Oxford: Oxford University Press, pp. 41–58.

Vallortigara, G. (2006c). The evolutionary psychology of left and right: Costs and benefits of lateralization. *Developmental Psychobiology*, **48**: 418–427.

Vallortigara, G. & Andrew, R. J. (1991). Lateralization of response to change in a model partner by chicks. *Animal Behaviour*, **41**: 187–194.

Vallortigara, G. & Andrew, R. J. (1994a). Differential involvement of right and left hemi-sphere in individual recognition in the domestic chick. *Behavioural Processes*, 33: 41–58.

Vallortigara, G. & Andrew, R. J. (1994b). Olfactory lateralization in the chick. *Neuropsychologia*, 32: 417–423.

Vallortigara, G. & Bisazza, A. (2002). How ancient is brain lateralisation? In: L. J. Rogers & R. J. Andrew (eds.), *Comparative Vertebrate Lateralization*, Cambridge: Cambridge University Press, pp. 9–69.

Vallortigara, G. & Rogers, L. J. (2005). Survival with an asymmetrical brain: Advantages and disadvantages of cerebral lateralization. *Behavioral and Brain Sciences*, 28: 575–633.

Vallortigara, G., Chiandetti, C. & Sovrano, V. A. (2011). Brain asymmetry (animal). *Wiley Interdisciplinary Reviews: Cognitive Science*, 2: 146–157.

Vallortigara, G., Cozzutti, C., Tommasi, L. & Rogers, L. J. (2001). How birds use their eyes: Opposite left–right specialisation for the lateral and frontal visual hemifield in the domestic chick. *Current Biology*, 11: 29–33.

Vallortigara, G., Pagni, P. & Sovrano, V. A. (2004). Separate geometric and non-geometric modules for spatial reorientation: Evidence from a lopsided animal brain. *Journal of Cognitive Neuroscience*, 16: 390–400.

Vallortigara, G., Regolin, L., Bortolomiol, G. & Tommasi, L. (1996). Lateral asymmetries due to preferences in eye use during visual discrimination learning in chicks. *Behavioural Brain Research*, 74: 135–143.

Vallortigara, G., Rogers, L. J. & Bisazza, A. (1999). Possible evolutionary origins of cognitive brain lateralization. *Brain Research Reviews*, 30: 164–175.

Vallortigara, G., Rogers, L. J., Bisazza, A., Lippolis, G. & Robins, A. (1998). Complementary right and left hemifield use for predatory and agonistic behaviour in toads. *NeuroReport*, 9: 3341–3344.

Vallortigara, G., Snyder, A., Kaplan, G. *et al.* (2008). Are animals autistic savants? *PLoS Biology*, 6: 208–214.

van den Berg, F. E., Swinnen, S. P. & Wenderoth, N. (2010). Hemispheric asymmetries of the premotor cortex are task specific as revealed by disruptive TMS during bimanual versus unimanual movements. *Cerebral Cortex*, 20: 2842–2851.

van der Hoort, B., Gutyerstam, A. & Ehrsson, H. H. (2011). Being Barbie: The size of one's own body determines the perceived size of the world. *PLoS One*, 6: e20195.

van Dijck, J. P. & Fias, W. (2011). A working memory account of spatial-numerical associations. *Cognition*, 119: 114–119.

van Dooren, T. J., van Goor, H. A. & van Putten, M. (2010). Handedness and asymmetry in scale-eating cichlids: Antisymmetries of different strength. *Evolution*, 64: 2159–2165.

Vandenberg, L. N. & Levin, M. (2009). Perspectives and open problems in the early phases of left–right patterning. *Seminars in Cell and Developmental Biology*, 20: 456–463.

Vauclair, J. (2004). Lateralization of communicative signals in nonhuman primates and the hypothesis of the gestural origin of language. *Interaction Studies*, 5: 365–386.

Vauclair, J. & Meguerditchian, A. (2008). The gestural origin of language and its lateralization: Theory and data from studies in nonhuman primates. In: S. Kern, F. Gayraud & E. Marsico (eds.), *Emergence of Linguistic Abilities: From Gestures to Grammar*, Cambridge: Cambridge Scholars Publishing, pp. 43–59.

Ventolini, N., Ferrero, E. A., Sponza, S. et al. (2005). Laterality in the wild: Preferential hemifield use during predatory and sexual behaviour in the black-winged stilt. *Animal Behaviour*, **69**: 1077–1084.

Versace, E., Morgante, M., Pulina, G. & Vallortigara, G. (2007). Behavioural lateralization in sheep (*Ovis aries*). *Behavioural Brain Research*, **184**: 72–80.

Verstynen, T., Tierney, R., Urbanski, T. & Tang, A. (2001). Neonatal novelty exposure modulates hippocampal volumetric asymmetry in the rat. *NeuroReport*, **12**: 3019–3022.

Vigh-Teichmann, I., Korf, H. W., Nűrnbeyer, F. et al. (1983). Opsin-immunoreactive outer segments in the pineal and parapineal organs of the lamprey (*Lampetra fluviatilis*), the eel (*Anguilla anguilla*) and the rainbow trout (*Salmo gairdneri*). *Cell and Tissue Research*, **230**: 289–307.

Vingiano, W. (1991). Pseudoneglect on a cancellation task. *International Journal of Neuroscience*, **58**: 63–67.

von Economo, C. & Koskinas, G. (1925). *Die Cytoarchitektonik der Hirnrinde des erwachsenen Menschen*. Berlin: Springer.

Voyer, D., Bowes, A. & Snaggi, M. (2009). Response procedures and laterality effects in emotion recognition: Implications for models of dichotic listening. *Neuropsychologia*, **47**: 23–29.

Voyer, D., Voyer, S. & Bryden, M. P. (1995). Magnitude of sex differences in spatial abilities: A meta-analysis and consideration of critical variables. *Psychological Bulletin*, **117**: 250–270.

Wallman, J. & Pettigrew, J. D. (1985). Conjugate and disjunctive saccades in two avian species with contrasting oculomotor strategies. *Journal of Neuroscience*, **5**: 1418–1428.

Wang, X., Yand, J., Shu, H. et al. (2011). Left fusiform BOLD responses are inversely related to word-likeness in a one-back task. *NeuroImage*, **55**: 1346–1356.

Wanker, R., Sugama, Y. & Prinage, S. (2005). Vocal labelling of family members in spectacled parrotlets, *Forpus conspiculatus*. *Animal Behaviour*, **70**: 111–118.

Ward, R. & Collins, R. L. (1985). Brain size and shape in strongly and weakly lateralized mice. *Brain Research*, **328**: 243–249.

Waters, N. S. & Denenberg, V. H. (1994). Analysis of two measures of paw preference in a large population of inbred mice. *Behavioural Brain Research*, **63**: 195–204.

Watkins, J. A. S. (1999). Lateralisation of auditory learning and processing in the domestic chick (*Gallus gallus domesticus*). Unpublished D.Phil. thesis, University of Sussex.

Watson, N. V. & Kimura, D. (1989). Right-hand superiority for throwing but not for intercepting. *Neuropsychologia*, **27**: 1399–1414.

Webb, J. E. (1969). On the feeding and behaviour of the larva of *Branchiostoma lanceolatum*. *Marine Biology*, **3**: 58–72.

Webb, J. E. (1975). The distribution of amphioxus. *Symposia of the Zoological Society of London*. **36**: 179–212.

Weekes, N. Y., Zaidel, D. W. & Zaidel, E. (1995). Effects of sex and sex role attribution on the ear advantage in dichotic listening. *Neuropsychology*, **9**: 62–67.

Weiss, R. A. (2009). Apes, lice and prehistory. *Journal of Biology*, **8**: 20.

Wells, D. L. (2003). Lateralized behavior in the domestic dog. *Behavioral Processes*, **61**: 27–35.

Wells, D. L. & Millsopp, S. (2009). Lateralized behaviour in the domestic cat, *Felis silvestris catus. Animal Behaviour*, **78**: 537–541.

Weniger, G., Lange, C. & Irle, E. (2006). Abnormal size of the amygdala predicts impaired emotional memory in major depressive disorder. *Journal of Affective Disorders*, **94**: 219–229.

Wentworth, S. L. & Muntz, W. R. A. (1989). Asymmetries in the sense organs and central nervous system of the squid *Histioteuthis. Journal of Zoology*, **219**: 607–619.

Westerhausen, R. & Hugdahl, K. (2008). The corpus callosum in dichotic listening studies of hemispheric asymmetry: A review of clinical and experimental evidence. *Neuroscience and Biobehavioral Reviews*, **32**: 1044–1054.

Westin, L. (1998). The spawning migration of European silver eel (*Aguilla anguilla* L.), with special reference to stocked eels in the Baltic. *Fisheries Research*, **38**: 257–260.

Weyers, P., Milnik, A., Muller, C. & Pauli, P. (2006). How to choose a seat in theatres: Always sit on the right side? *Laterality*, **11**: 181–193.

Whiten, A. (2005). The second inheritance system of chimpanzees and humans. *Nature*, **437**: 52–54.

Whiten, A., Goodall, J., McGrew, W. C. *et al.* (2001). Charting cultural variation in chimpanzees. *Behaviour*, **138**: 1481–1516.

Whiten, A., Schick, K. & Toth, N. (2009). The evolution and cultural transmission of percussive technology: Integrating evidence from palaeoanthropology and primatology. *Journal of Human Evolution*, **57**: 420–435.

Wichman, A., Freire, R. & Rogers, L. J. (2009). Light exposure during incubation and social and vigilance behaviour in domestic chicks. *Laterality*, **14**: 381–394.

Wichman, A., Rogers, L. J. & Freire, R. (2008). Visual lateralization and development of spatial and social spacing behaviour of chicks (*Gallus gallus domesticus*). *Behavioural Processes*, **81**: 14–19.

Wild, B., Rodden, F. A., Grodd, W. *et al.* (2003). Neural correlates of laughter and humour. *Brain*, **126**: 2121–2138.

Wild, J. M., Williams, M. N. & Suthers, R. A. (2000). Neural pathways for bilateral vocal control in songbirds. *Journal of Comparative Neurology*, **423**: 413–426.

Wiltschko, W. & Wiltschko, R. (2005). Magnetic orientation and magnetoreception in birds and other animals. *Journal of Comparative Physiology A*, **191**: 675–693.

Wiltschko, W. & Wiltschko, R. (2009). Avian navigation. *The Auk*, **126**: 717–743.

Wiltschko, W., Munro, U., Ford, H. & Wiltschko, R. (2003). Lateralization of magnetic compass orientation in silvereye, *Zosterops lateralis. Australian Journal of Zoology*, **51**: 1–6.

Wiltschko, W., Traudt, J., Güntürkün, O., Prior, H. & Wiltschko, R. (2002). Lateralization of magnetic compass orientation in a migratory bird. *Nature*, **419**: 467–470.

Wiltschko, W., Wiltschko, R. & Ritz, T. (2011). The mechanism of the avian magnetic compass. *Procedia Chemistry*, **3**: 276–284.

Wisniewsky, A. B. (1998). Sexually-dimorphic patterns of cortical asymmetry, and the role for sex steroids in determining cortical patterns of lateralisation. *Psychoneuroendocrinology*, **23**: 519–547.

Witelson, S. F. (1976). Sex and the single hemisphere: Specialization of the right hemisphere for spatial processing. *Science*, **193**: 425–427.

Witelson, S. F. & Nowakowski, R. S. (1991). Left out axons make men right: A hypothesis for the origin of handedness and functional asymmetry. *Neuropsychologia*, **29**: 327–333.

Witelson, S. F., Kigar, D. L., Scamvougeras, A. *et al.* (2008). Corpus callosum anatomy in right-handed homosexual and heterosexual men. *Archives of Sexual Behavior*, **37**: 857–863.

Workman, L. & Andrew, R. J. (1989). Simultaneous changes in behaviour and in lateralization during the development of male and female domestic chicks. *Animal Behaviour*, **38**: 596–605.

Yamazaki, Y., Aust, U., Huber, L., Hausmann, M. & Güntürkün, O. (2007). Lateralized cognition: Asymmetrical and complementary strategies of pigeons during discrimination of the 'human concept'. *Cognition*, **104**: 315–344.

Yáněz, J., Busch, J., Anadón, R. & Meissl, H. (2009). Pineal projections in the zebrafish (*Danio rerio*): Overlap with retinal and cerebellar projections. *Neuroscience*, **164**: 1712–1720.

Yáněz, J., Pombal, M. A. & Anadón, R. (1999). Afferent and efferent connections of the parapineal organ in lampreys: A tract tracing and immunocytochemical study. *Journal of Comparative Neurology*, **403**: 171–189.

Young, J. Z. (1962). Why do we have two brains? In: V. B. Mountcastle (ed.), *Interhemispheric Relations and Cerebral Dominance*, Baltimore, MD: Johns Hopkins Press, pp. 7–24.

Yu, D., Akalal, D.-B. G. & Davis, R. L. (2006). Drosophila α/β mushroom body neurons form a branch-specific, long-term cellular memory trace after spaced olfactory conditioning. *Neuron*, **52**: 845–855.

Zalc, B., Goujet, D. & Colman, D. (2008). The origin of the myelination programme in vertebrates. *Current Biology*, **18**: R511–R512.

Zappia, J. V. & Rogers, L. J. (1983). Light experience during development affects asymmetry of fore-brain function in chickens. *Developmental Brain Research*, **11**: 93–106.

Zeigler, H. P. & Marler, P. (2008). *Neuroscience of Birdsong*. Cambridge: Cambridge University Press.

Zeitlin, S. B., Lane, R. D., O'Leary, D. S. & Schrift, M. J. (1989). Interhemispheric transfer deficit and alexithymia. *American Journal of Psychiatry*, **146**: 1434–1439.

Zhuralev, A. V. (2007). Morphofunctional analysis of late Paleozoic conodont elements and apparatuses. *Paleolontical Journal*, **41**: 549–557.

Zucca, P. & Sovrano, V. A. (2008). Animal lateralization and social recognition: quails use their left visual hemifield when approaching a companion and their right visual hemifield when approaching a stranger. *Cortex*, **44**: 13–20.

Zucca, P., Baciadonna, L., Masci, S. & Mariscoli, M. (2011a). Illness as a source of variation of laterality in lions (*Panthera leo*). *Laterality*, **16**: 356–366.

Zucca, P., Cerri, F., Carluccio, A. & Baciadonna, L. (2011b). Space availability influences laterality in donkeys (*Equus asinus*). *Behavioural Processes*, **88**: 63–66.

# Index

Abe, 79
abstraction, 166
Adamec, 13, 73, 141
Adelstein, 75
Ades, 94
adrenalin, 155
advantages/disadvantages of asymmetry, 39–41, 49, 51
Agetsuma, 71, 101, 127
aggression, 14, 21, 24–25, 27, 39–40, 52, 54, 60, 72, 73, 92, 94, 104, 106–107, 110, 116, 118–119, 130–131, 138, 145, 154–155, 157–158, 163, 166
agonistic behaviour. *See* aggression
Aizawa, 71, 101, 127
Albert, 20, 116
Albertson, 60
*Alectura lathami*, 110
alexithymia, 161
Aljuhanay, 75
Alkonyi, 126, 134
Allan, 79
Allman, 16, 136–137
Almécija, 82
Alonso, 17, 117
altricial species, 108–109
Alvarez, 131
Alves, 32, 92
Alzheimer's disease, 150
*Amatitlania nigrofasciata*. *See* cichlid fish
amodal completion, 21
amphibians, 13, 24–25, 56. *See also* toad
*Amphioxus*. *See* lancelet
amusement, 130, 138, 140
amygdala, 23, 73, 130–131, 136, 138–139, 141, 147–149, 151–152, 165
Anderson, 157
Andrew, 4–8, 13, 22, 24, 26–27, 30, 32, 42, 47, 68, 73, 75, 77, 84–85, 97,

100–102, 118, 120, 127, 130, 138, 144, 150
androgens. *See* sex hormones
Anfora, 30, 58, 59, 94
anger, 74, 131
animate targets, 10, 14
annelids, 33, 89
Annett, 9, 52
Anokhin, 151
*Anolis carolinensis*. *See* lizard
Anson, 3, 4, 7
ant, 30, 94
antenna, 28, 30, 58–59, 94–97, 170
antibodies. *See* immune responses
antisymmetry, 50
anxiety, 13, 165. *See also* fear
apes. *See* primates
*Apis mellifera*. *See* honeybee
arachnids. *See* spider
*Araneae*. *See* spider
Arbib, 80
arrow worms, 68–69
Artelle, 157
Arthropoda, 62, 89–90, 92, 94–97
Asami, 31
Asperger's syndrome, 149
*Asymmetron*. *See* lancelet
attention, 27, 128, 132–133, 136, 167
audition, 4, 27, 78, 84, 124, 133, 147, 164
Austin, 14, 73, 154, 160, 163, 166
*Australopithecus afarensis*. *See* hominid
*Australopithecus sediba*. *See* hominid
autism, 137, 149, 162
autonomic nervous system, 16, 27
axis, left–right, 89–90

Babcock, 94
baboon, 10, 14, 26, 39, 73, 76, 85, 88–89, 166
Baguñà, 90

Baldwin, 23
Balzeau, 82
Bambach, 62
Banzan, 131
Barbalet, 148
Barca, 140
Baron-Cohen, 149
bat, 87
Bateson, 4, 122, 159
Beaumont, 38
bee-eater, 111
*Belostoma flumineum. See* water bug
beluga whale, 27
Bennett, 74, 159
Berezinskaja, 69
Berlim, 161
Berrebi, 117
Bertram, 87
Bianco, 71, 127
bilateral symmetry, 35, 65
Bilateria, 33
Billiard, 75
Binkofski, 80, 83
bipedal locomotion, 82
birds, 3, 12–13, 17, 25–26, 87, 100, 103,
     110–111, 114, 142, 158, 170, 206
Bisazza, 12–13, 25, 26, 44, 50–52, 56–57, 74,
     140
Bitan, 147
Blanke, 135
body hair, 88
body image, 134, 139
Boere, 157
boldness, 74, 100, 111, 159
Boleda, 8
Boles, 48
*Bombus* spp. *See* bumblebee
Bonati, 74, 154, 163, 165
Bone, 68
Bonetti, 78
bonobo, 9, 81–82
Booker, 95
Boorman, 33, 90
Booth, 149
Borod, 15
bottom-up, 20
Boughman, 87
Boycott, 31
Braccini, 74, 159
*Brachyraphis episcope. See* fish
Bradley, 104

Bradshaw, 22, 38, 102, 114, 121
Brain, 20
brain efficiency, 41, 50–51
brain's capacity. *See* neural capacity
Braitenberg, 2, 37, 102
*Branchiostoma. See* lancelet
Branson, 160
Breedlove, 7
Broca, 1, 8, 10, 81, 83–84
Broca's area, 80, 83, 85, 89, 134–136, 140, 147,
     149
Brown, 48, 58, 102
Brownell, 140
Brunt, 94
brush turkey, 110
Buccino, 80, 83
Buckner, 137–138, 150
Budaev, 100–101
Budil, 94
*Bufo bufo. See* toad
*Bufo marinus. See* toad
bumblebee, 30
Burgdorf, 139
Burghardt, 143
Buzsáki, 128, 138
Byers, 142
Byrne, 8, 92

Cabeza, 136, 150
*Caenorhabditis elegans. See* nematode
Caine, 74, 159
Callaert, 135
Cameron, 74, 159
Cammarota, 151
Campbell, 2
canary, 39, 78–79, 115
*Canis familiaris. See* dog
Canli, 140, 148
Cantalupo, 10, 42, 82, 128
*Carduelis tristis. See* goldfinch
Caron, 63
Carrasquillo, 157
Casey, 12, 141
Casperd, 14, 27, 73, 166
Castelli, 149
cat, 11, 13, 141, 157
categorization, 17, 23, 46–47, 121, 123,
     131–132, 134, 149, 162, 164, 166
cephalopods, 92
cerebellum, 81, 105, 128
cerebral torque, 82

chaetognaths. *See* arrow worms
Chapman, 161
Charron, 45, 133, 144
Chen, 1, 64
Cherkin, 5
Chi, 135, 167
Chiandetti, 7, 14, 21, 107
chick, 3, 4–7, 12–14, 18–22, 24, 39, 42–45, 47,
    73, 75–77, 84, 100, 103–107, 109–110,
    115–120, 128–129, 131, 144–145, 150,
    163–164
chimpanzee, 8–10, 48–49, 74, 76, 81–89, 128,
    138, 141, 159
chirality, 31
chordates, 62–64, 66, 68–69, 90, 97
Chura, 118, 146
cichlid fish, 2, 3
cingulate cortex, 15, 136–137, 147
Cipolla-Neto, 5, 30
Clark, 72, 117
claw asymmetry, 95
Clayton, 30, 75, 99
cockatoo. *See* parrot
cockroach, 94
cognitive bias, 158–159
Collins, 50, 145
Colonnese, 78
colour perception, 121, 131–132
commissure, 143
computation, 46, 48–49, 52
concept formation, 17
Concha, 71, 101
configuration, 17
Cooper, 95
cooperative interactions, 57
Corballis, 8, 22, 52, 82, 164
cornutes, 64–65
corpus callosum, 38, 114, 116, 118, 120,
    143–144, 146–147, 161
cortical gyrification, 10
cortical thickness, 126
corticosterone. *See* stress hormones
corvid, 142
counting, 76
cowbird, 79
Cowell, 75, 114
Craig, 15, 137, 163
Cristino, 102
Crockford, 85
Crow, 161
Crowne, 75

Csermely, 154, 165
C-start reaction, 40–41, 51
cuttlefish, 32, 89
cyclostomes. *See* jawless fish
*Cymatogaster aggregate. See* perch

Da Costa, 23
da Guardia, 128
Dadda, 41, 44
Daisley, 24
*Danio rerio. See* zebrafish
dark incubation. *See* darkness
darkness, 44, 84, 100, 104, 107, 109, 111, 133
Davidoff, 121
Davidson, 15, 73, 130
Davison, 31
de Boer, 88
de Boyer des Roches, 114
de Gelder, 131
de Latude, 74
De Santi, 26–27, 140
decision-making, 16, 27, 45, 83, 133
Deckel, 25, 73, 206
decussation, 6, 37, 40, 68, 73, 94, 108, 124
deduction, 135
default network, 136–137, 150
defensive reflex, 37
Dehaene-Lambertz, 121
*Delphinapterus leucas. See* beluga whale
dementia, 162
Denenberg, 3, 11, 50, 114, 117, 158
Deng, 24, 104, 109–110, 115, 129
Denny, 157, 161
depression, 15, 149, 157–158, 161–162
Deruelle, 76
deuterostomes, 33, 89–90
Dharmaretnam, 119
Diamond, 126
Diba, 128, 138
dichotic listening, 124, 143
Diekamp, 17, 20, 144
Dien, 123
diencephalon. *See* midbrain
Dimond, 15, 73
direction of lateralization, 50–51, 53, 79, 102,
    163
directional lateralization, 57–59
discrimination, 4, 6, 10, 13, 20, 23, 48,
    84, 121
disgust, 73, 130
distraction, 27, 125, 127, 145, 149

distress, 73, 131, 157
Diver, 31
divergent thinking, 135, 167
dog, 11, 12, 15, 16, 22, 73, 78, 87, 125,
    154–156, 160
dolphin, 27, 86, 163
dominance, of right eye system, 109
Dong, 68
Donoghue, 68
dorsal network, 133–134, 143, 150
Doupe, 79
Dowling, 70
Downhill, 13
Downs, 149
Drach, 64
Drews, 25, 73
*Drosophila melanogaster. See* fruitfly
dual task, 44
Duguid, 95
Duistermars, 30, 95
Dunbar, 14, 27, 73, 166
dunnart, 13, 166
Dzemidzic, 8

eagle, 142
Eaton, 76
Ecdysozoa, 33, 89, 90
echinoderms, 62–65
eel, 72, 102
eggshell, 12, 78, 103
Ehret, 84, 124
Ehrlichman, 73
Ehrman, 9
elephant, 16, 137, 157
Emberley, 76
embryo, 12, 78, 99–100, 102–104, 106–107, 109,
    110, 114–116, 122
emergency responses, 83, 157–158, 166
emotion, 13–16, 22–23, 27, 72–74, 78, 85,
    123–124, 126–127, 129–132, 136–139, 141,
    143, 147–149, 151, 153–156, 161, 163, 165,
    167–168, 170
empathy, 136–137, 149, 166–168
Engbretson, 71, 102
Enggist-Dueblin, 87
episodic memory, 47, 145, 150
Erwin, 25
Esslinger, 161
evolutionarily stable strategy, 52, 55–57,
    61, 168
experience, 99

face, 22–24, 26, 130, 168–169
face perception, 22, 26, 28, 75, 140, 150
face touching, 15
facial expression, 15, 131, 168
Fagot, 10, 76
Fan, 128, 137
Faurie, 9, 55, 75
fear, 13, 16, 27, 71–74, 107, 130, 141, 154,
    156–159. *See* anxiety
Ferbinteau, 128
Fernald, 92
Fernandez-Carriba, 15, 22
Ferrari, 86
Fias, 132
Ficken, 142
Field, L., 3
Finch, 8
finch, 79
fish, 10, 12–13, 18, 24–27, 34, 38–41, 44, 51,
    57–60, 63, 68–70, 72, 74, 76, 80, 90, 100–102,
    111, 140, 143, 158, 165
Fitch, 117
flamingo, 157
Foà, 71
Folta, 19, 21
food caching, 75, 98
food webs, 62
foot preference, 1, 9, 12
foraging, 13, 30, 48, 61, 87, 94,
    97, 129
*Formicidae. See* ant
Forrester, 10
Foster, 53
Foundas, 146
Fox, 133–134, 135
Franklin, 121, 132
Frasnelli, 28, 89, 170
Freake, 71
Frederick, 73, 131
Fredes, 129
Freire, 115, 119–120
Frith, 149
frontal cortex, 23, 45, 74
fruitfly, 30, 32, 90, 95–96, 170
Fu, 4
Fuqua, 25
fusiform gyrus, 139–140

*Gambusia holbrooki. See* fish
Gainotti, 138
Galaburda, 116

Gallate, 168
*Gallus gallus. See* chick
game theoretical model, 52
Gardner, 87
gastropods, 90
Gazzaniga, 38, 150
Geng, 134
Gentilucci, 80
geometry, 17–18, 75, 128
George, 3
gerbil, 117, 139
Gereau, 157
Geschwind, 116
gesture, 10
gesturing, 9, 10, 83, 85–86, 89
Ghirlanda, 52–54, 169
Gianotti, 141
Gibbs, 5, 6, 151
Gilbert, 122, 132
Gilissen, 82
Giljov, 15, 68
*Girardinus falcatus. See* fish
global cues, 76
Gluckman, 122
goldfinch, 86
Goldstein, 15
Gomez, 67
Goodwin, 112
Gordon, 159
gorilla, 8, 10, 15, 81–82, 88
*Gorilla gorilla. See* gorilla
Gorrie, 10, 82
Goto, 127
Goulson, 30
Govind, 95
Grace, 163
Grande, 33, 90
Gray, 139
greeting, 85–86
Greicius, 137
Grimm, 157, 160–161
Groothuis, 116
Gruber, 129
Guenther, 4
Guglielmotti, 102
Guiard, 80
Guioli, 103
gull, 116
Güntürkün, 17, 74, 77, 108–110, 129
Gutiérrez-Ibánezfoun, 3
Gutnick, 92

Haakonsson, 157
Haase, 30
Habas, 128–129
habenula, 2, 3, 70–73, 100–102, 126–127
Häberling, 146
*Haikouella*, 64–66, 68
Hall, 168
Halpern, 101–102
Hamilton, 7, 22
Hampson, 146
hand control of, 82
hand preference, 10–11, 48–49, 74, 80–83,
     159–161
  and homicide, 55
  and sport, 55, 75
hand, control of, 80, 82–83, 85, 147
handedness, 2, 8–10, 12, 39, 48, 52, 62, 75, 79,
     82–83, 116, 126, 128, 146, 157, 161, 164. *See
     also* hand preference
handling, 54, 74, 99, 113–114, 117, 158, 160
Hanley, 121
haptic discrimination, 10
Hardyck, 9
Harmon-Jones, 74, 131, 138
Harries, 15
Harris, 2, 9
Harvey, 88
hawk, 142
Hazlerigg, 67
head turning, 39, 41, 72, 78, 154, 156, 162, 168
hearing. *See* audition
Hecht, 148
*Helix. See* snail
Hellige, 8, 112
hemichordates, 90
hemineglect, 18, 20
hemispheric communication, 147
hemispheric dominance, shifts in, 118, 120–121
hemispheric interaction, 124, 143–144, 151, 167
Herlitz, 146
Heuts, 30, 40, 94
Hewes, 83
Hews, 14, 24–25
Hickok, 135
higher vocal centre, 3, 78
Higuchia, 83
Hikosaka, 127
*Himantopus himantopus. See* stilt
hippocampus, 18–19, 72, 92, 98–99, 114,
     126–128, 136–138, 150–152, 161, 165
Hirnstein, 48

*Histioteuthis*, 92
Hobert, 33
Hochner, 92
Hodos, 2
Hoff, 76
Hoffman, 163
Holdstock, 128
Holland, 67, 135
homicide, 55, 75
hominid, 82, 88
*Homo heidelbergensis*, 88
homosexual/heterosexual difference, 148
honeybee, 28, 58, 94, 96, 170
Hook, 10, 159
Hopkins, 8, 10, 15, 22, 48, 74, 76, 81–82, 84, 128, 159
Hopp, 86
hormone. *See* stress hormones; sex hormones
Horn, 4, 150
Horowitz, 87
horse, 14, 39, 73, 114, 154, 155–156, 160–161, 166
Hourcade, 97
Howard, 121
Hox genes, 63
Hugdahl, 78, 143
Hui-Di, 139
human, 1–4, 7–10, 13–16, 18, 22–23, 38, 42, 44–46, 55, 61–62, 64, 73–76, 78–89, 98, 111–112, 116, 118, 120–121, 126, 128, 130–131, 134, 136–141, 144, 147, 150, 154–156, 160–162, 164, 166, 168, 208
human foetus, 111, 118
Hyafil, 133
Hymenoptera, 30, 58, 189
hyoid bone, 88
hypoglossal nerve, 3

Iacoboni, 136
immune responses, 11
imprinting, 4, 22, 119
inanimate targets, 10, 14
Ingle, 76
interhemispheric communication, 146–147, 157, 166
intermediate medial mesopallium, 5, 24
interpeduncular nucleus, 71–72, 101
invariance, 47
Iturria-Medina, 126
Izquierdo, 151

Jacobs, 98
James, 145
Jamieson, 70
jawed fish, 70
jawless fish, 2
Jefferies, 64
Jennings, 73
Johanson, 144
Johnson, 98
Johnston, 150
Joseph, 149
Juraska, 117

Kanwisher, 22
Kaplan, 15, 79, 81, 87, 145, 160, 163, 168
Kappers, 127
Karenina, 27
Kawakami, 127
Keenan, 4
Kells, 30
Kemali, 2, 102
Kendrick, 23
Kesh, 17
Kight, 31, 94
Kilpatrick, 147
Kim, 11, 117, 150, 209
Kimura, 80, 145–147
King, 80
Kipper, 139
kissing direction, 39, 74
Kiuchi, 111
Koboroff, 141, 163
Kocot, 92
Koechlin, 45, 133, 144
Kon, 63
kookaburra, 163
Koskinas, 15
Kosslyn, 75, 131
Kovach, 103
Krebs, 75, 99, 170
Kuan, 101–102
Kuhl, 79
Kwok, 121

Lacalli, 66–67, 70
Lamon, 138
lamprey, 2, 71
lancelet, 63–64, 66–70
Land, 92
Landau, 80
landmark, 19, 43, 44, 119

Lane, 73
Langford, 167
language, 2–3, 7–10, 38, 42, 62, 79–83, 87–89,
    97, 116, 121, 126, 134, 146, 165, 170
large paired neurons, 67
Larose, 154
Lartillot, 95
*Larus ridibundus*. *See* gull
*Lasius niger*. *See* ant
laughing, 15, 139
laughter, 138–139
Laviola, 142
Lavrysen, 82, 144
learning, 4–6, 20, 24, 29–30, 44, 79, 84–86, 90,
    92, 94–95, 100, 119–120, 122, 127–128,
    132, 133, 150–151, 200
*Leipoa ocellata*. *See* mallee fowl
Leith, 72
Letzkus, 28–29, 30, 58, 94
Levin, 89, 103
Levy, 23, 42
Lewis, 64
Li, 66
light exposure, 44, 100–102, 104, 106–107, 109,
    110–111, 115, 158
*Limax*. *See* slug
limb preference, 8, 12, 27, 94, 158–161
Lindström, 148
Linnoila, 25
lion, 160, 161
Lippolis, 13, 25, 40, 68, 73, 166
lizard, 14, 24, 25, 39, 71–73, 102, 130, 163, 165
Llorente, 8
lobster, 95
Lonsdorf, 8, 48, 81
Lössner, 5
Loudon, 67
Love, 115
Lovell-Badge, 103
Lowe, 90
Lüders, 126
lungfish, 13, 68
Lurito, 8
*Lymnaea stagnalis*. *See* snail

*Macaca mulatta*. *See* primate; rhesus monkey
macaque, 139
MacNeilage, 8, 10, 13, 15, 36, 79, 80, 83, 112, 166
MacPherson, 133
Magat, 48
magnetic compass, 163

magpie, 39, 79, 141, 163
Maguire, 18, 98, 150
Maillard, 140
Malakhov, 69
Malaschichev, 12
Mallatt, 64, 68
mallee fowl, 110
mammals, 10, 13, 18, 26–27, 67, 71–72, 77–78,
    83, 111, 116–118, 120, 124, 126–127, 129,
    130, 143, 151, 165, 170
mangabeys. *See* primates
Mangun, 134
manipulation, 13, 83
    of objects, 10
Manns, 129
Marchant, 8, 9, 48, 49
Mari, 149
Marler, 79
marmosets, 10, 74, 82, 157, 159, 177
marsh tit, 75, 98
Marshall, 84, 85
marsupial, 15, 142
Martin, 77, 137, 139, 157, 163
Marzoli, 39
mason bee, 58
Matheson, 159
Matsusaka, 138
Mauthner cells, 40, 42, 76
Maynard-Smith, 52
Mazzotti, 157
McCourt, 75
McGilchrist, 160, 162, 166
McGinnis, 139
McGreevy, 156, 160
McGrew, 8, 48–49
McKenzie, 24, 73, 154
McManus, 8, 9, 50, 52, 55, 164
megapode, 110
Meguerditchian, 8–10, 85
Mehlhorn, 99, 109
memory, 5–6, 29, 46, 98, 127, 133, 135, 146, 161
memory formation, i, ix, 4–6, 30, 92, 96–97,
    101, 137, 150–152, 170, 173
memory recall, 5, 17, 29, 30, 32, 46, 58, 94, 96,
    97, 147, 150–152, 170
Mench, 144
Mendl, 159
Merckelbach, 75
Messenger, 92
Michel, 112
midbrain, 2, 3, 71, 77, 102, 108, 127, 129

migration, 163
Miklósi, 68, 74, 84, 127, 141, 165
Miller, 150, 162
mimicry, 79
Minagawa-Kawai, 121
mirror neurons, 80, 83, 87, 136
mirror test, 26
Mishkin, 84, 133
Mitchell, 162
Miyasaki, 73
Mobbs, 139
mollusc, 89–90, 92
mood, 138
Morris, 22, 75, 130, 139
motor cortex, 10, 83, 128, 147
mouse, 10–11, 50, 51, 78, 83, 142, 157
mouth asymmetry, 59, 64
Mulckhuyse, 130
Müller-Oehring, 143–144
Mummery, 150
Muntz, 92
mynah, 87
Myowa-Yamakoshi, 86

Nagy, 26
Narang, 78, 117
*Nautilus*, 92
navigation, 17, 18, 99
Nematoda, 95
nematode, 33–34, 65
Nepi, 58
nesting behaviour, 110
Nestor, 14
Nettleton, 22
neural capacity, 42, 44
Nicholls, 168
Niemitz, 157
Nir, 82, 84
Nixon, 32
Nodal genes, 33, 90, 103
noise phobia, 160
Nottebohm, 3, 78–79
Nowakowski, 118
Nowicka, 126, 143, 146
Nowicka and Tacikowski, 146
*Nucifraga columbiana. See* nutcracker
nucleus accumbens, 139
nucleus rotundus, 19–20, 77, 105, 108
Núñez, 117
Nunn, 88
nutcracker, 21, 98

Ocklenburg, 74
octopus, 92
oestrus, 88
Ohlmann, 145
Okubo, 39
Oldfield, 9
Oldowan culture, 82
olfaction, 27–30, 34, 36–37, 58, 72, 73, 77, 95,
    97, 125, 155, 164
Olko, 125
opposable thumb, 81
optic nerves, 6, 78, 117
optic tectum, 77, 108–109, 129
orang-utan, 15, 81, 88, 168
orbitofrontal cortex, 148, 165
order of laying, 115–116
originality, 135
*Orrorin. See* hominid
Ortiz, 95
ovenbird, 111

pain perception, 157
paired sense organs, 62, 66–67
Palmer, 8, 50, 60
*Pan troglodytes. See* chimpanzee
Panganiban, 90
Panksepp, 139
Papadatou, 146
Papademetriou, 8
*Papio cynocephalus. See* primate
parapineal, 70–72, 100–102
parasympathetic. *See* autonomic nervous
    system
parietal, 31, 71, 133–136, 147
parietal cortex, 75, 136
parietal eye, 71
Park, 150
Parker, 160–161
Parr, 22
parrot, 1, 9, 48, 79, 86–88
Pascual, 32, 96
passive avoidance learning, 5, 150–151
Pasteels, 78
Patel, 33, 90
Paterson, 62
pathology, 38, 149
Patterson, 6
paw preference, 11–12, 50–51
pawedness. *See* paw preference
pebble floor task. *See* pebble–grain test
pebble–grain test, 7, 17, 44, 107

Pecchia, 27
Peirce, 23
Pellis, 143
Pepperberg, 87
perch, 41
Perelle, 9
*Periplaneta americana. See* cockroach
*Perissodus microlepis. See* fish
personality, 164
Peters, 81
Pettigrew, 77
Pfister, 87
Phelps, 131
Philippe, 95
Phillips, 73, 131, 154, 191, 202
photoreceptors, 33, 37, 46–47, 66–67, 70, 117, 163
Piekema, 128
Pierson, 78
pigeon, 17, 19–21, 23, 26–27, 48, 77, 99, 100, 108–110, 129, 144
Pinsk, 22
Pisella, 136
placoderms. *See* jawed fish
planum parietale, 146
planum temporale, 84
play, 87, 130, 139–140, 142, 159
pointing, 80–81, 85
Poirier, 3
*Pongo pygmaeus. See* orang-utan
Pontier, 9
population lateralization, 30, 35, 50–53, 55–58, 60–61
Poremba, 84, 133
Prather, 87
predator, attention to, 13, 26–27, 44, 140, 159
predator, response to, 27, 39, 51, 53, 57–58, 60, 68, 73–74, 94, 101, 107
prefrontal cortex, 132–136, 138–139, 141, 148–150, 152, 161
prejudice, 168
premotor cortex, 82, 133, 135
Previc, 112
prey capturing, 13, 21, 51, 67–68, 72, 76, 92, 129, 142, 159
primates, 8, 10, 15–16, 22, 25–26, 74, 78–80, 82–85, 87–88, 112, 128–130, 137, 139, 159, 160
proboscis extension response, 29, 58
Proverbio, 140
pseudoneglect, 20
psychopath, 148, 165

psychosis, 161
Pu, 70
Puzdrowski, 129

quail, 17, 24
Quaranta, 11, 15–16, 73, 156, 160
Quinn, 95

radial symmetry, 35
Rahman, 148
Raichle, 137
Rajendra, 116
Ramirez, 94
Ramon Y Cajal, 37
Rashid, 75
rat, 3, 11, 13, 18, 25, 39, 72–73, 75, 78, 98, 99, 113–114, 116–118, 126–128, 131, 138–139, 141–142, 167
raven, 87
Raymond, 9, 55, 75
reaching, 8, 11, 81, 85, 149, 159, 160
Reeb, 158
Regier, 132
Regolin, 22
releasing stimuli, 73
Ren, 132, 164
reptiles, 24, 64, 70, 154
reptilian ancestors, 18
Reverberi, 135, 137
reversal of asymmetry, 146
reversed asymmetry, 52, 104, 132
rhesus monkey, 22, 82, 84, 117
Rickard, 6
Rieger, 69
Rigosi, 30, 94
risk taking, 137, 140–142, 166
Rizzolatti, 80
Roberson, 121
Robert, 145
Roberts, 70
robin, 163
Robins, 13–14, 25, 68, 73, 154
Robinson, 13, 138
Robison, 94
rodent, 3, 18, *See also* rat, mouse
Rogers, 3, 4, 7–10, 12–15, 24–25, 28–29, 33, 44–45, 52, 58, 68, 73–74, 81, 94, 97, 99, 102–104, 107, 109–110, 113–116, 118–120, 122, 129–130, 145, 154, 156–160, 163–165, 168, 170, 173
Rosa, 4, 22, 24
Rose, 4–5, 150, 173

Rota-Stabelli, 92
routine behaviour, 15, 27, 39
Rowe, 77
Rugani, 21, 76
Rugg, 150
Ryan, 117

saccades, 77, 136
Sackeim, 15
Safer, 14
Sagasti, 95
Saint-Galli, 139
Sakai, 27
Salva, 22, 24
Samara, 18
Sandi, 6
Sandoz, 95–96
Santrock, 8
savant, 162, 167
Savic, 148
Saykin, 128
Sceloporus virgatus. See lizard
Schachter, 46
Schenker, 83
Schiff, 138
Schmidt, 3
schizophrenia, 137, 148, 161
Schmahmann, 128
Schomerus, 63
Schulte, 143–144
Schwabl, 115–116
Schwarz, 116
Scytodes globosa. See spider
Seeck, 23
Seeger, 139
Seeley, 136–137
Seger, 132
Semendeferi, 137
Semple, 157
sensitive period, 100–102, 106, 107, 111,
    115–116, 119, 121–122
Sepia. See squid
Sepia officinalis. See cuttlefish
Sepioteuthis. See squid
serotonin, 25
sex difference, 11, 98, 116–118, 126, 128, 140,
    144–148
sex hormones, 106, 116–118, 141,
    144–146, 149
sexual behaviour, 24–25, 39, 72–74, 104,
    106–107, 118–119, 130–131, 145, 163

Shallice, 128
Shamay-Tsoori, 135
Shapiro, 128
Shapleski, 84
Sharp, 72
Shaw, 161
Shaywitz, 147
sheep, 22–23, 26, 154
Sherwood, 10, 83
Shettleworth, 98
Shimeld, 33, 90
Shin, 157, 160
Shomrat, 92
Shu, 70
Shulman, 134
Siegel, 104
silvereye, 163
Sindhurakar, 104
Siniscalchi, 125, 154–155, 160, 163
Sink, 104
Siok, 121
slug, 90
Smaers, 134, 138
Smart, 117
Smith, 63, 149
Smotherman, 114
snail, 31, 33, 89–90
Snyder, 135, 162, 167
social behaviour, 16, 22, 25, 47, 58, 83, 86, 136,
    148, 163
social cognition, 17, 24, 27–28
social displays, 163
social structure, 88, 130
song, 3
    of birds, 3–4, 78–79
songbird, 3
Sovrano, 24–26, 73
sparrow, 87
spatial cognition, 6, 10, 17–19, 25, 28, 75, 98,
    119, 128
Spear, 87, 141
speech, 3, 10, 42, 78–84, 88, 97, 121, 128,
    134–135, 146–147, 150
Spencer, 98
Sperry, 7
spider, 30, 94
Spivak, 161
split-brain patients, 7, 38, 143
Spocter, 10, 83–84
Spreng, 137
squid, 92–93

starling, 115
Steklis, 8
steroid hormones, 115
Stewart, 60
stilt, 39, 163
Stokes, 64, 67
Stoodley, 128
strength of lateralization, 48, 50, 51, 58, 102, 122, 149
stress, 13, 48, 56, 85, 99, 114–115, 117, 153, 157–158, 160–161
stress hormones, 115
stroke, 73, 162
*Sturnus vulgaris. See* starling
Sullivan, 158
Summerfield, 133
Summers, 151
superior colliculus, 77, 130
Surace, 78
sustained response, 70, 80, 101, 165
Sutherland, 72
Suthers, 79
Swalla, 63
swimming, 32, 63–64, 66–70
Sylvian fissure, 81, 82
sympathetic. *See* autonomic nervous system
synergistic interactions, 54
syrinx, 3, 79
systematizing, 149
Szaniawski, 68

Tacikowski, 126, 143
tactile stimulation, 114
*Taenopygia guttata. See* zebra finch
Tager-Flusberg, 149
Taglialatela, 85
tail wagging, 15–16, 156
Tan, 11
Tang, 158
targets, 144
Taube, 72
Taylor, 34
tectofugal system, 20, 77, 108–109
Telford, 92
temporal lobe, 135, 165, 167, 168
temporal sulcus, 22, 165
termite fishing, 48, 81
tern, 163
testosterone. *See* sex hormones
thalamofugal system, 77, 109
Thatcher, 120

Theeuwes, 130
*Theropithecus gelada. See* baboon
Thomas, 74
throwing, 8–9, 80
toad, 12–14, 21, 25, 39, 52, 56, 66, 68, 70, 73, 130, 165
Todt, 139
Tomaz, 157
Tomer, 89
Tommasi, 6, 12, 18, 39, 42, 68, 75–77, 119, 128
tool use, 9–10, 82–83
top-down, 20, 133
touching, 147
Tourville, 4
Town, 24
Tranel, 136
transcranial magnetic stimulation, 167
Treherne, 53
*Trichogaster trichopterus. See* fish
trilobite, 65, 94
Tsakiris, 134
Tucker, 73, 131
Tully, 95
Tulogdi, 72
Tulving, 150
Turkewitz, 125
turning, 11, 12, 41, 57, 68, 74, 94, 125, 154, 160
*Tursiops aduncus. See* dolphin
turtle, 143
two task, 45–46, 133, 144

unpaired motor organs, 42
*Urosaurus ornatus. See* lizard

Vakulenko, 139
valence hypothesis, 15
Valencia-Alfonso, 109, 129
Valenti, 17
Vallortigara, 6–8, 12–14, 17–18, 21–22, 24–25, 27, 29, 33, 41–43, 46–47, 52–53, 57–58, 68, 73, 75–76, 94, 97, 107, 119, 162–163, 165, 169–170
van den Berg, 82
van der Hoort, 134
van Dijck, 132
Van Dooren, 60
Vandenberg, 89
Vandenbergh, 117
variance, 27, 46, 47
Vauclair, 8, 9, 10

Ventolini, 25, 163
ventral network, 133–136, 141–142
Vermeire, 22
Versace, 154
Verstynen, 114, 158
Vigh-Teichmann, 71
visual field, 13, 20–22, 24, 38–39, 41, 73, 76–77,
    124, 129, 131–132, 134, 136, 140, 145,
    154, 175
visual pigment, 67
visual system, 77, 78, 107. See also thalamofugal
    system; tectofugal system; optic tectum;
    superior colliculus
vocal apparatus, 42, 83, 85, 88
vocalization, 79, 86
von Economo, 15
von Economo neurons, 15–16, 137
Voyer, 144–145

Wallman, 77
Wang, 140
Wanker, 86
Ward, 51
Wassersug, 12
Watanabe, 79
water bug, 31, 94
Waters, 11
Watkins, 27, 84
Watson, 80
Weiss, 88
welfare of animals, 153, 160–161
wellbeing, 157
Wells, 11, 160
Weniger, 149
Wentworth, 92

Wernicke's area, 10, 81–82, 84,
    134–135
Westin, 72
Weyers, 39
whale, 16, 64, 137
Whiten, 81–82
Whorfian hypothesis. See colour perception
Wichman, 44, 107
Wild, 32, 79, 139
Wilson, 71, 127, 148
Wiltschko, 163
Wisneiwski, 117
Wisniewsky, 143, 147
Witelson, 118, 147–148
Workman, 9, 12, 115, 118, 120, 163
Worthington, 14, 25

Xenopus. See toad

Yamazaki, 17, 23
Yáněz, 70, 71, 102
Young, 32, 37
Youngren, 73, 131
Yu, 96

Zalc, 70
Zappia, 104
zebra finch, 3, 4, 17
zebrafish, 27, 34, 56, 71–73, 100, 102–103, 127,
    151, 170
Zeigler, 79
Zeitlin, 161
zero-sum game, 41
Zhuralev, 68
Zucca, 24, 160–161

Printed in the United States
by Baker & Taylor Publisher Services

Printed in the United States
by Baker & Taylor Publisher Services